Johannes Schinko

Analysis of Tribolium head patterning

Johannes Schinko

Analysis of Tribolium head patterning

by forward and revers genetics and transgenic techniques

Südwestdeutscher Verlag für Hochschulschriften

Impressum/Imprint (nur für Deutschland/ only for Germany)
Bibliografische Information der Deutschen Nationalbibliothek: Die Deutsche Nationalbibliothek verzeichnet diese Publikation in der Deutschen Nationalbibliografie; detaillierte bibliografische Daten sind im Internet über http://dnb.d-nb.de abrufbar.
 Alle in diesem Buch genannten Marken und Produktnamen unterliegen warenzeichen-, marken- oder patentrechtlichem Schutz bzw. sind Warenzeichen oder eingetragene Warenzeichen der jeweiligen Inhaber. Die Wiedergabe von Marken, Produktnamen, Gebrauchsnamen, Handelsnamen, Warenbezeichnungen u.s.w. in diesem Werk berechtigt auch ohne besondere Kennzeichnung nicht zu der Annahme, dass solche Namen im Sinne der Warenzeichen- und Markenschutzgesetzgebung als frei zu betrachten wären und daher von jedermann benutzt werden dürften.

Verlag: Südwestdeutscher Verlag für Hochschulschriften Aktiengesellschaft & Co. KG
Dudweiler Landstr. 99, 66123 Saarbrücken, Deutschland
Telefon +49 681 37 20 271-1, Telefax +49 681 37 20 271-0
Email: info@svh-verlag.de
Zugl.: Göttingen, Universität, Dissertation, 2009

Herstellung in Deutschland:
Schaltungsdienst Lange o.H.G., Berlin
Books on Demand GmbH, Norderstedt
Reha GmbH, Saarbrücken
Amazon Distribution GmbH, Leipzig
ISBN: 978-3-8381-1465-1

Imprint (only for USA, GB)
Bibliographic information published by the Deutsche Nationalbibliothek: The Deutsche Nationalbibliothek lists this publication in the Deutsche Nationalbibliografie; detailed bibliographic data are available in the Internet at http://dnb.d-nb.de.
 Any brand names and product names mentioned in this book are subject to trademark, brand or patent protection and are trademarks or registered trademarks of their respective holders. The use of brand names, product names, common names, trade names, product descriptions etc. even without a particular marking in this works is in no way to be construed to mean that such names may be regarded as unrestricted in respect of trademark and brand protection legislation and could thus be used by anyone.

Publisher: Südwestdeutscher Verlag für Hochschulschriften Aktiengesellschaft & Co. KG
Dudweiler Landstr. 99, 66123 Saarbrücken, Germany
Phone +49 681 37 20 271-1, Fax +49 681 37 20 271-0
Email: info@svh-verlag.de

Printed in the U.S.A.
Printed in the U.K. by (see last page)
ISBN: 978-3-8381-1465-1

Copyright © 2010 by the author and Südwestdeutscher Verlag für Hochschulschriften Aktiengesellschaft & Co. KG and licensors
All rights reserved. Saarbrücken 2010

Table of Content

1 Summary **3**

2 General Introduction **5**

    2.1   Insect head patterning: *Drosophila* 5

    2.2   The Red Flour Beetle: A model system with insect typical head 6

    2.3   The first large scale insertional mutagenesis screen in *Tribolium* 7

    2.4   Aims of this thesis 9

        2.4.1   Functional analysis of head gap-like gene orthologs 9

        2.4.2   Insertional mutagenesis Screen and analysis of mutants 9

        2.4.3   Establishment of misexpression techniques 10

3 Results **12**

    3.1   Divergent functions of *orthodenticle*, *empty spiracles* and *buttonhead* in early head patterning of the beetle *Tribolium castaneum* (Coleoptera) 13

    3.2   Large-scale insertional mutagenesis of the coleopteran stored grain pest, the red flour beetle *Tribolium castaneum*, identifies embryonic lethal mutations and enhancer traps 28

        3.2.1   Manuscript 29

        3.2.2   Insertion site analysis of lines generated in Göttingen 60

        3.2.3   Rescreen of selected lethal lines 62

            3.2.3.1   Analysis of lethal line G02408 63

            3.2.3.2   Analysis of lethal line G07411 65

            3.2.3.3   Analysis of lethal line G07521 66

            3.2.3.4   Analysis of lethal line G09104 68

            3.2.3.5   Analysis of lethal line G10215 70

            3.2.3.6   Analysis of lethal line KT1269 72

            3.2.3.7   Analysis of lethal line KS0294 76

            3.2.3.8   Discussion of lethal screen 78

        3.2.4   Rescreen of selected enhancer trap lines 81

            3.2.4.1   Enhancer trap line G03920 82

	3.2.4.2	Enhancer trap line G10011 84
	3.2.4.3	Enhancer trap line G11122 86
	3.2.4.4	Discussion of enhancer trap screen.. 87

    3.3   The binary expression system GAL4/UAS in *Tribolium* 90

4   ***General Discussion*** ***117***

5   ***References*** ***119***

6   ***Appendix*** ***125***

    6.1   Supplementary Table 126

    6.2   Sequences of rescreen of lethal lines... 127

    6.3   Sequences of GAL4/UAS constructs 143

    6.4   Abbreviations 157

1 Summary

In this thesis several different methods have been applied to get deeper insights into the complex process of head development in insects. As model organism the red flour beetle *Tribolium castaneum* was chosen. This is a well suited organism for analyzing this process, as – in contrast to *Drosophila* – *Tribolium* larvae exhibit an insect typical head with all head appendages.

First, I established a map of head bristles that serve as landmarks for head defects.

Second, I analyzed the so called head gap-like genes *orthodenticle, empty spiracles* and *buttonhead* in *Tribolium*. These genes are known to play a crucial role in *Drosophila* head patterning. I analyzed the expression pattern and the phenotype caused by knock down via RNA interference and compared this to *Drosophila* data. I find that depending on dsRNA injection time, two functions of *Tc-orthodenticle1* can be distinguished. The early regionalization function affects all segments formed during the blastoderm stage while the later head patterning function is similar to *Drosophila*. In contrast, both expression and function of *Tc-empty spiracles* are restricted to the posterior part of the ocular and the anterior part of the antennal segment and *Tc-buttonhead* is not required for head cuticle formation at all. I conclude that the gap gene like roles of *ems* and *btd* are not conserved while at least the head patterning function of *otd* appears to be similar in fly and beetle.

Third, I set out to identify novel genes involved in head development because by the reverse genetics approach genes that are not known to be involved in head patterning in other organisms will not be found. I screened the 2612 insertion lines generated in the Göttingen part of the insertional mutagenesis screen for enhancer traps and analyzed the cuticle phenotype of all lethal lines. The insertion site of the mutator was determined and assigned to the probably affected gene. Seven interesting lines of the 328 (238 from Göttingen and 90 from Kansas) lethal lines were analyzed in more detail. This led to three very interesting lines affecting head development that will be matter of future analysis.

Summary

In order to analyze gene function in more detail, not only knock down but also misexpression is needed. Hence, the fourth aim of this thesis was to establish the binary expression system GAL4/UAS in *Tribolium*. GAL4 is activated via heat shock by using the *Tribolium hsp68* promoter. Upon heat shock the reporter gene is expressed via the UAS construct in all stages of *Tribolium* development and different tissues as well. This work also revealed that it is essential to make use of *Tribolium* specific basal promoters in the GAL4 as well as in the UAS construct.

2 General Introduction

2.1 Insect head patterning: *Drosophila*

The structure of the insect head with regard to both its ontogeny and evolution has been debated controversely for decades. Whereas classical morphological data mainly take into account the structure of the nervous system, coelom cavities and head appendages (Rempel 1975), recent approaches consider molecular and genetic data like expression patterns or knock-out phenotypes (Schmidt-Ott and Technau 1992; Rogers and Kaufman 1997). However, the number of head segments and the potential contribution of non-segmental tissue are still a matter of debate (Rempel 1975; Scholtz and Edgecombe 2006).

So far, both morphological and embryological molecular data argue for the subdivision of the insect head into three pregnathal (ocular, antennal and intercalary) and three gnathal (mandibular, maxillary and labial) head segments that are assigned to the mouth parts (Rempel 1975; Jürgens 1986; Cohen and Jurgens 1991). The debate on the character of the pre-ocular region, sometimes referred to as the unsegmented acron, and the character of the labrum has remained highly controversial even in the light of molecular data and is still ongoing (Haas et al. 2001; Haas et al. 2001; Scholtz and Edgecombe 2006) (Posnien, in press.). If the number of head segments is determined by the adjacent *wingless/engrailed* expression domains, an additional labral segment exists in *Drosophila* (Schmidt-Ott and Technau 1992). However, this is not found in most arthropods (Scholtz and Edgecombe 2006). Recently it has been argued based on data from *Tribolium* that the labrum is part of non-segmental tissue (Posnien, in press.)

Most molecular data on developmental processes in insects is undoubtfully available for the fruit fly *Drosophila melanogaster*. The gnathal segments contribute to the posterior part of the head and are, except for the mandibular segment (Vincent et al. 1997), established like the trunk segments via the genetic cascade of maternal determinants, gap-, pair rule- and segment polarity genes (Cohen and Jurgens 1990; Johnston 1992). In parallel homeotic genes are activated that establish segment identity (McGinnis and

Krumlauf 1992). The mandibular segment gets input from both the trunk cascade and the anterior patterning system.

In the pregnathal region, downstream of the maternal determinants, the head gap-like genes *orthodenticle* (*otd*), *empty spiracles* (*ems*) and *buttonhead* (*btd*) are activated to metamerize the *Drosophila* head (Cohen and Jurgens 1990; Cohen and Jurgens 1991). It has remained unknown if these functions are conserved among insects or are an adaption to the special mode of head development in *Drosophila* (see chapter 3.1 for more details).

2.2 The Red Flour Beetle: A model system with insect typical head

While many important insights have been gathered in this species, *Drosophila* is not best suited for the analysis of insect head patterning. On the one hand, this is due to the phylogenetic position of *Drosophila* within insects that suggests that it represents a highly derived state of development. Furthermore, the larval head is invaginated into the thoracic segments. This head involution causes a derived larval head morphology and leads to difficulties in analyzing head phenotypes especially if head involution defects occur which elicit secondary defects. Moreover, markers for distinct head regions are sparse (Dalton et al. 1989; Nassif et al. 1998). Therefore the lab uses the red flour beetle *Tribolium castaneum* to gain insight into the process of head development. *Tribolium* belongs to the order of the Coleoptera that comprises 350.000 species and hence is the species-richest order on earth. In contrast to *Drosophila*, *Tribolium* larvae possess an insect-typical head including all insect-specific head appendages (Bucher and Wimmer 2005).

Apart from these properties that render *Tribolium* a suitable organism to analyze head development, its technical accessibility makes it an insect model system second only to *Drosophila*. Stock keeping is easy with ample offspring all year round and a fast generation time of about four weeks. The genome is sequenced (Richards et al. 2008), robust RNAi techniques are established (Brown et al. 1999; Bucher et al. 2002; Tomoyasu and Denell 2004; Miller et al. 2008) and RNAi has been shown to be splice-variant-specific (Arakane et al. 2005). Germ line transformation in *Tribolium* is as efficient as in *Drosophila* using a universal marker system (Berghammer et al. 1999; Horn et al. 2002).

Gene knock down via RNAi is straightforward in *Tribolium* but gain of function analysis has not been possible. In *Drosophila*, misexpression tools have enabled the deep analysis of many processes. The binary GAL4/UAS system is one example of a very useful and versatile tool for targeted gene expression (Brand and Perrimon 1993). In short, the yeast originated GAL4 is a transactivator that binds to an Upstream Activating Sequence (UAS) and thereby activates expression of genes downstream of UAS. This system has been adapted successfully to several species like mouse, zebrafish, *Arabidopsis* and *Xenopus* (Ornitz et al. 1991; Guyer et al. 1998; Scheer and Campos-Ortega 1999; Hartley et al. 2002).

Despite several trials during the previous five years (Bucher, Klingler, Wimmer, pers. comm.) no misexpression system has been available for *Tribolium* although the genetic toolbox has expanded enormously in the last decade.

2.3 The first large scale insertional mutagenesis screen in *Tribolium*

In order to learn about the genes involved in development, various homologous genes were isolated from *Tribolium* and their expression and function were compared to their *Drosophila* orthologs (reverse genetics) (Angelini et al. 2009; Economou and Telford 2009; Parthasarathy and Palli 2009; Posnien 2009; Yang et al. 2009). However, due to technical difficulties in *Drosophila* (see above), no comprehensive list of head patterning genes exists. Moreover, this approach cannot uncover genes that are crucial for pattern formation in the beetle but not in the fly or other organisms. In order to identify such genes small scale chemical mutagenesis screens have been performed in *Tribolium* (Sulston and Anderson 1996; Maderspacher et al. 1998). The genome of *Tribolium castaneum* consists of nine autosomal chromosome pairs and two gonosomes. Only less than one third of the genome is covered by balancers (Trauner et al., submitted.). Thus stock keeping of unmarked recessive mutations is difficult (Berghammer et al. 1999).

These problems can be circumvented by an insertional mutagenesis screen. A system to realize such a screen was established recently (Lorenzen et al. 2007). By the use of a dominantly marked "donor" transposon both stock keeping and gene identification is facilitated. Hence, the GEKU consortium (**G**öttingen –Ernst Wimmer; **E**rlangen - Martin

Klingler, **K**ansas State University - Sue Brown; **U**SDA grain marketing and production research center - Richard Beeman) initiated the **"GEKU"** insertional mutagenesis screen in 2005. This is the first large-scale insertional mutagenesis screen conducted in an insect other than *Drosophila*.

2.4 Aims of this thesis

2.4.1 Functional analysis of head gap-like gene orthologs

I applied several different methods to unravel mechanisms of insect head development. First, I conducted a candidate gene approach. An advantage of this reverse genetic approach is that results are obtained quickly as gene knock-down via RNAi is efficiently working in *Tribolium* and cuticles can subsequently be analyzed for head-specific phenotypes. I analyzed the function of the head gap-like genes *orthodenticle* (*otd*), *empty spiracles* (*ems*) and *buttonhead* (*btd*) that are known to play a crucial early role in head development of *Drosophila*. The question was if their function is conserved in *Tribolium* or if their role in *Drosophila* is an adaptation to its derived mode of head development. Moreover, the head bristle pattern of *Tribolium* had to be described in order to have a framework for the interpretation of head phenotypes (see published manuscript "Divergent functions of *orthodenticle, empty spiracles* and *buttonhead* in early head patterning of the beetle *Tribolium* castaneum (*Coleoptera*)" in chapter 3.1).

2.4.2 Insertional mutagenesis Screen and analysis of mutants

An unbiased way to identify genes involved in certain developmental processes is a mutagenesis screen (forward genetics). Random mutagenesis, either caused by chemical mutagens or by insertional mutagenesis can affect any gene and therefore identify new genes important for the process of interest. In order to do so, the mutated lines have to be screened for phenotypes and the affected gene has to be determined. In case of an insertional mutagenesis the integration site of the mutator can for example be determined by performing inverse PCR (iPCR) (Ochman et al. 1988). In many cases this information will help to assign a phenotype to a gene. A large scale insertional mutagenesis screen was conducted by the GEKU consortium in *Tribolium* (Trauner et al., submitted) and I screened the lines for enhancer traps in the head and other tissues and for mutations affecting head development (see submitted manuscript "Large-scale insertional mutagenesis of the coleopteran stored grain pest, the red flour beetle *Tribolium castaneum*, identifies embryonic lethal mutations and enhancer traps in

chapter 3.2.1). My task within this collaborative effort was to screen all 2612 lines generated in Göttingen for embryonic, larval, pupal and adult enhancer traps. I also determined the genomic integration site of most lethal and sterile insertions generated in Göttingen as well as the integrations of the most interesting enhancer traps via inverse PCR. Then, the genomic locations, enhancer-trap patterns (if present), recessive phenotypes, and genes affected by these transposon insertions were documented in the GEKU database (available at www.geku-base.uni-goettingen.de). Moreover, I screened all Göttingen and Kansas lethal lines for head defects and other defects of the L1 cuticle (see 3.2). Some of the data are not part of the manuscript and are provided in additional chapters:

3.2.2 "Insertion site analysis of lines generated in the GEKU screen in Göttingen",
3.2.3 "Rescreen of selected lethal lines",
3.2.4 "Rescreen of selected enhancer trap lines".

2.4.3 Establishment of misexpression techniques

In order to analyze the function of genes involved in head patterning in detail not only knock-down but also over- and misexpression of these genes is useful. As *Tribolium* is a relatively young model organism compared to *Drosophila*, transgenic techniques are still sparse. Since binary expression systems have been adapted to multiple uses in *Drosophila* (Duffy 2002) it is of key interest to establish such a technique in *Tribolium*. Previous attempts to establish the GAL4/UAS system were based on the use of the *Drosophila* constructs including the *Drosophila* basal promoter. I hypothesized that a potentially decreased activity of a *Drosophila* basal promoter in *Tribolium* could have been the reason for the negative results, so I chose to use an endogenous promoter from *Tribolium* for the constructs. To adapt the system to *Tribolium* I used both GAL4Δ and GAL4-VP16. GAL4Δ is a truncated version of GAL4 where the DNA binding domain is directly fused to the transactivation domain (Ma and Ptashne 1987). In the version GAL4-VP16 the activation domain of GAL4 was replaced by the stronger activation domain of the herpes simplex virus protein VP16 (McKnight et al. 1987; O'Hare and Goding 1988; O'Hare et al. 1988; Preston et al. 1988; Triezenberg et al. 1988; Triezenberg et al. 1988). Both versions worked in *Tribolium* and I found evidence that indeed endogenous promoters are crucial for the success (see 3.3: Manuscript in preparation "The binary expression system GAL4/UAS in *Tribolium*")

3 Results

Every chapter within the results starts with a one-page description of:

- the main aim of the particular manuscript in the context of the thesis
- the authors and their contributions to the practical work, and
- the status of the manuscript.

3.1 Divergent functions of *orthodenticle*, *empty spiracles* and *buttonhead* in early head patterning of the beetle *Tribolium castaneum* (Coleoptera)

In this part, the so called head gap-like genes, which play an important role in *Drosophila* head development, have been analyzed in *Tribolium*. Their expression pattern and knock-down phenotype was compared to *Drosophila*. In addition the bristle pattern of the larval head has been described in detail and serves as a landmark for mapping patterning defects.

Johannes B. Schinko, Nina Kreuzer, Nils Offen, Nico Posnien, Ernst A. Wimmer, Gregor Bucher

Authors contributions to the practical work:

J.B. Schinko:	All practical work except for:
Nina Kreuzer:	Analysis of *Tc-ems* and *Tc-btd* expression pattern; initial RNAi experiments with *Tc'ems* and *Tc-btd*.
Nils Offen:	Cloning of *Tc-btd* gene
Nico Posnien:	Analysis of head bristle pattern of 10 wt larvae

Status: Published in Developmental Biology (Elsevier) 317 (2008); pp.: 600–613

Developmental Biology

journal homepage: www.elsevier.com/developmentalbiology

Divergent functions of *orthodenticle*, *empty spiracles* and *buttonhead* in early head patterning of the beetle *Tribolium castaneum* (Coleoptera)

Johannes B. Schinko [a], Nina Kreuzer [a], Nils Offen [b], Nico Posnien [a], Ernst A. Wimmer [a], Gregor Bucher [a,*]

[a] *Department of Developmental Biology, Johann Friedrich Blumenbach Institute of Zoology and Anthropology, Georg-August-University Göttingen, Germany*
[b] *Department of Biology, Evolutionary Biology, University of Konstanz, Germany*

ARTICLE INFO

Article history:
Received for publication 17 December 2007
Revised 3 March 2008
Accepted 4 March 2008
Available online 15 March 2008

Keywords:
Arthropod head
Insect head
Head gap gene
Vertex
Gena
Brain
Orthodenticle
Empty spiracles
Buttonhead
SP5
Segmentation

ABSTRACT

The head gap genes *orthodenticle* (*otd*), *empty spiracles* (*ems*) and *buttonhead* (*btd*) are required for metamerization and segment specification in *Drosophila*. We asked whether the function of their orthologs is conserved in the red flour beetle *Tribolium castaneum* which in contrast to *Drosophila* develops its larval head in a way typical for insects. We find that depending on dsRNA injection time, two functions of *Tc-orthodenticle1* (*Tc-otd1*) can be identified. The early regionalization function affects all segments formed during the blastoderm stage while the later head patterning function is similar to *Drosophila*. In contrast, both expression and function of *Tc-empty spiracles* (*Tc-ems*) are restricted to the posterior part of the ocular and the anterior part of the antennal segment and *Tc-buttonhead* (*Tc-btd*) is not required for head cuticle formation at all. We conclude that the gap gene like roles of *ems* and *btd* are not conserved while at least the head patterning function of *otd* appears to be similar in fly and beetle. Hence, the ancestral mode of insect head segmentation remains to be discovered. With this work, we establish *Tribolium* as a model system for arthropod head development that does not suffer from the *Drosophila* specific problems like head involution and strongly reduced head structures.

© 2008 Elsevier Inc. All rights reserved.

Introduction

The arthropod head is built by several segment primordia that fuse to form the rigid head capsule with limbs that are adapted to different roles in feeding (Snodgrass, 1935; Weber, 1966). Curiously, the number of head segments and the potential contribution of non-segmental tissue (Rempel, 1975; Scholtz and Edgecombe, 2006) as well as the evolution of the labrum remain disputed (Budd, 2002; Haas et al., 2001). Also, the developmental genetics of insect head patterning remain enigmatic because head involution of *Drosophila melanogaster* leads to a derived and reduced larval head morphology that has hampered mutant analysis (Fig. 1) (Nassif et al., 1998; Dalton et al., 1989; Jürgens et al., 1986).

The gnathal segments (mandible, maxilla and labium) constitute the posterior portion of the arthropod head. Except for the mandible (Vincent et al., 1997), they are patterned like the other trunk segments in *Drosophila* according to the well established hierarchical segmentation cascade involving maternal coordinate genes, gap-, pair rule- and segment polarity genes (St Johnston and Nusslein-Volhard, 1992). Segment identity specification is accomplished by the Hox genes (McGinnis and Krumlauf, 1992). Anterior to the mandible, however, no pair rule patterning is observed and the anterior most expression of a

* Corresponding author. Fax: +49 551 395416.
E-mail address: gbucher1@uni-goettingen.de (G. Bucher).

0012-1606/$ – see front matter © 2008 Elsevier Inc. All rights reserved.
doi:10.1016/j.ydbio.2008.03.005

gene of the Hox cluster is in the intercalary segment (Bucher and Wimmer, 2005; Diederich et al., 1991). For patterning of anterior head segments, *Drosophila* makes use of the so-called head gap like genes *otd*, *ems* and *btd* (Cohen and Jürgens, 1991; Cohen and Jürgens, 1990). They are expressed early in embryogenesis in broad and overlapping domains (Dalton et al., 1989; Finkelstein and Perrimon, 1990; Walldorf and Gehring, 1992; Wimmer et al., 1993, 1997). Also their deletion domains are overlapping and correspond roughly to the expression patterns. As judged by cuticular phenotypes as well as *engrailed* and *wingless* (*wg*) expression in mutants, loss of *otd* function leads to a loss of the ocular and the antennal segments. *ems* is required for the posterior most parts of the ocular segment as marked by the absence of the *engrailed* head spot but presence of the adjacent ocular *wg* domain (head blob). In addition, *ems* mutants lack antennal and intercalary segments (Cohen and Jürgens, 1990) and the anterior portion of the mandibular segment (Dalton et al., 1989; Walldorf and Gehring, 1992). Mutations in *btd* lead to loss of antennal, intercalary, mandibular and the anterior portion of the maxillary segments (Wimmer et al., 1996). The head gap genes are not required for patterning the labrum (Cohen and Jürgens, 1990, 1991; Wimmer et al., 1996). Their regulation depends on three maternal systems: Via *bicoid* the anterior system activates *otd*, *ems* and *btd*, the terminal system is required for anterior positioning of both *otd* and *btd* and the dorso-ventral system provides further pattern refinement (Dalton et al.,

14

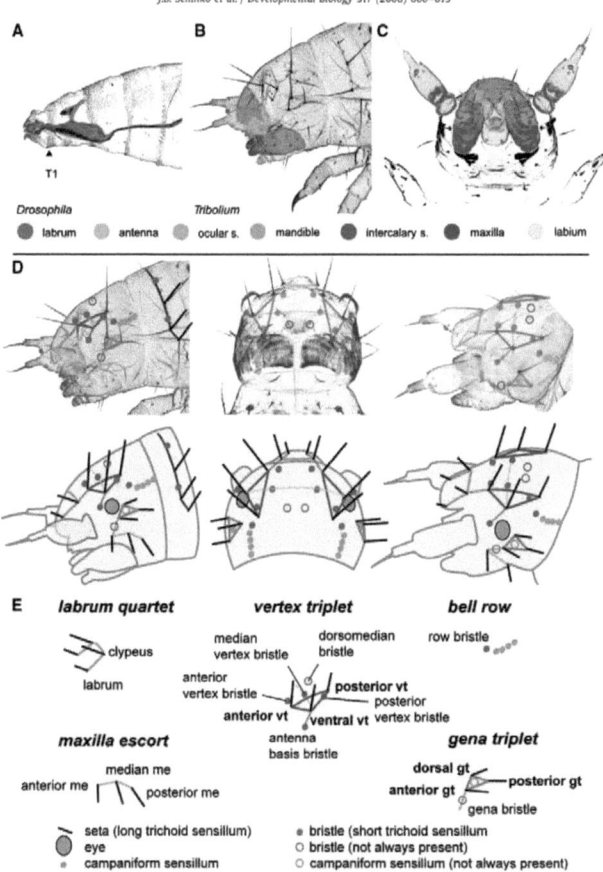

Fig. 1. Head cuticles of Drosophila and Tribolium. (A) The Drosophila larval head is involuted into the thorax and its structures are heavily reduced. (B, C) The Tribolium larva in contrast displays a head with all structures typical for insect heads (B: lateral view; C: ventral view; color code for head appendages is the same in A–C). (D) The dorsal and lateral portions of the Tribolium head are marked by a pattern of setae and bristles. Groups of setae are connected by colored lines. The angles of the colored lines mark the position of the long setae while red dots indicate short bristles. Orange dots indicate campaniform sensillae while open circles mark bristles/campaniform sensillae that are not found on all wild-type cuticles. (E) For future reference, setae and bristles have been grouped and been given names. Note that neither color code nor names indicate any developmental or segmental units. See text for details.

1989; Finkelstein and Perrimon, 1990; Gao and Finkelstein, 1998; Gao et al., 1996; Grossniklaus et al., 1994; Wimmer et al., 1995). This activation by maternal factors together with the gap phenotypes has led to their classification as gap like genes. The suggestion that their combinatorial action would also specify segmental identity (Cohen and Jürgens, 1991; Grossniklaus et al., 1994) has not proven correct (Gallitano-Mendel and Finkelstein, 1998; Wimmer et al., 1997). ems but not otd has some homeotic selector function and the gap phenotypes btd to do so (Gallitano-Mendel and Finkelstein, 1998; Schöck et al., 2000).

Strikingly, in vertebrates the orthologs of otd and ems are expressed in the anterior brain Anlage and are required for its development (Acampora et al., 1998; Reichert and Simeone, 1999; Simeone et al., 1992; Treichel et al., 2003; Wimmer et al., 1993). Cross-phylum experiments have shown that the murine Otx and Emx2 proteins are able to partially rescue respective Drosophila mutant brain phenotypes (Hartmann et al., 2000; Leuzinger et al., 1998). The Drosophila Otd protein in turn has an activity similar to the endogenous ortholog in Xenopus (Lunardi and Vignali, 2006) and mouse brain development (Acampora et al., 1998). Together with other data, this has led to the view of an urbilaterian origin of the animal brain and of highly conserved brain patterning mechanisms (Denes et al., 2007; Lichtneckert and Reichert, 2005; Reichert and

Simeone, 2001; Tessmar-Raible et al., 2007). However, the murine SP factor that had previously been described as buttonhead (mBtd) (Treichel et al., 2003) actually belongs to the SP8 family (Beermann et al., 2004; Griesel et al., 2006). Hence, the striking functional similarities of murine mBtd and Drosophila btd became arguable while the true vertebrate ortholog of Dm-btd has remained unclear.

With respect to head development, Drosophila is a poor representative of the arthropods because of its highly derived mode of larval head development. In contrast to Drosophila, the red flour beetle Tribolium develops a regular larval head with all structures typical for an insect head (Bucher and Wimmer, 2005). With robust RNAi techniques established (Brown et al., 1999; Bucher et al., 2002; Tomoyasu and Denell, 2004) and with the genome sequenced (The Tribolium Genome Sequencing Consortium, in press) Tribolium has evolved into an arthropod model system second only to Drosophila.

With this work, we introduce Tribolium as a model for larval head development in arthropods. First we describe the bristle pattern of the larval head in order to provide landmarks for mapping patterning defects. Then we analyze the orthologs of the Drosophila head gap like genes Tc-otd1, Tc-ems and Tc-btd. We find that the latter two do not function as head gap genes and that changes in gene function correlate with altered expression patterns. Our data suggest that cross-phyla comparisons of gene function should not be based on highly variable early patterning processes.

Materials and methods

Phylogenetic analysis

The zinc-finger and buttonhead boxes of Tc-SP8, Drosophila and mouse SP-factors were aligned using ClustalW with subsequent manual curation. Only clearly aligned positions were used for the phylogenetic analysis. A phylogenetic tree was calculated using the Tree-Puzzle algorithm (Schmidt et al., 2002) at http://bioweb.pasteur.fr/seqanal/interfaces/Puzzle.html with standard options but "calculation of clock-like branch lengths" and the "more exact algorithm". The resulting tree was visualized using TreeView 1.6.6. (R.D.M. Page 2001). To identify additional conserved motifs, the mouse Sp5, Dm-Btd and Tc-Btd proteins were subjected to pair wise dot blot analysis at SRS with standard settings. The subsequent ClustalW alignment as implemented in the MEGA 3.1 (Kumar, Tamura, Nei) did not properly align all domains identified in the dot blots and were hence curated manually as well as the N-terminal part of the alignment. The alignment was displayed by "boxshade" at www.ch.embnet.org/software/BOX_form.html. Mouse Emx1/2 and Dm-Ems were retrieved from the SRS protein database. tBlastn analysis using Tc-Ems as query in the Drosophila genome revealed the ems, e5 and the ex ex genes as best hits. Blast of Dm-E5 in Tribolium retrieved Tc-Ems, Tc-Bagpipe and Tc-Exex. Pair wise dot blot analysis and ClustalW alignments, calculation and representation of the tree and alignment depiction were done as described above. As Mm-Emx1 and Mm-Emx2 appeared very similar in the dot blot analysis, only Mm-Emx1 was used for the analysis.

Double stainings

False color representation of double fluorescence images was performed as described in (Wohlfrom et al., 2006). Note that the NBT/BCIP staining quenches the fluorescent fastRed staining to some extent such that weak overlaps may be missed.

RNAi

Templates were prepared by PCR with T7-primers from plasmid template comprising full length (1.1 kb) plus 125 bp 5'UTR and 700 bp 3'UTR of the Tc-otd1 and 520 bp of the open reading frame plus 200 bp 3'UTR of Tc-ems. DsRNA was produced using the Megascript Kit (Ambion). Concentrations for parental RNAi were 2–4 μg/μl (Tc-otd1), 2.5 μg/μl (Tc-ems) and 1–5 μg/μl (Tc-btd) and for embryonic RNAi 0.1–1 μg/μl (Tc-otd1 and Tc-btd) and 0.5 μg/μl (Tc-ems). Injections were performed as described (Brown et al., 1999; Bucher et al., 2002). To test for the portion of embryos that do not develop cuticle, we injected 3.7 μg/μl Tc-otd1-dsRNA and collected eggs of injected and non-injected animals three times, respectively. Of both hatching and non-hatching larvae, cuticles were prepared. The number of eggs in the egg collection was counted as well as the portion of developed cuticles and empty eggshells without visible remnants of cuticles (wt: n=238, Tc-otd1RNAi: n(d7)=31 n(d8)=45 n(d10)=106). The experiment was done three times using two independently cloned Tc-otd1 templates.

Microscopy

Cuticle preparations were documented by laser scanning microscopy as described before (Wohlfrom et al., 2006). For the presentation, the colors have been inverted using Photoshop 7.0 (Adobe). Nomarski optics and fluorescent images of the whole mount in situ stainings were documented using a Zeiss Axioplan microscope (Zeiss, Jena).

Testing negative results for Tc-btd

Pupae were injected with 5 μg/μl dsRNA comprising the entire coding region (756 bp). A small portion of the egg collection (7 days after injection) was used for cuticle preparations, the rest was fixed. Whole mount in situ staining against Tc-btd including Tc-caudal as positive control in the same color reaction revealed no detectable Tc-btd activity but normal Tc-caudal expression in most embryos. All cuticles of this egg collection were wild type and the portion of empty egg shells was normal. To test for potential later Tc-btd function, we also allowed eggs of injected pupae to develop. The ratio of hatching larvae was within the normal range (6 of 19 injected as compared to 10 of 21 in the buffer control) and all hatched larvae developed to morphologically normal adult beetles. Also injection of dsRNA into embryos did not cause overt phenotypes.

Results

The Tribolium L1 larval head

For better interpretation of cuticular phenotypes, we have determined a set of cuticular structures that mark different regions of the Tribolium head. The insect head is a complex structure that arose by the fusion of the proposed acron–an anterior non-segmental tissue (i.e. not serial homologous to trunk segments)–and several segments that are serial homologous to trunk segments. Some authors suggest that the acron is minuscule or even absent (Schmidt-Ott and Technau, 1992). The Tribolium larva is prognath, i.e. the mouth opening is oriented toward anterior. The mouth appendages encircle the mouth opening forming a preoral cavity. This "ring" is closed posteriorly by the labium and laterally by mandibles and maxillae. These gnathal appendages are markers for the posterior portion of the respective segments as they arise from Tc-engrailed/Tc-wingless (Tc-wg) positive tissues. The antennae are oriented toward anterior and are a marker for the posterior portion of the antennal segment. The larval eyes are located posterior to the antennae with respect to the larval anterior–posterior-axis (ap-axis). They reside below the cuticle and their position is not marked by any specific cuticular structure. Before clearing of cuticles, however, they can be identified within the head capsule and used as markers for part of the ocular region. The clypeolabrum is located between the two antennae and projects downward to encircle the preoral cavity from the front. The basis of the clypeus probably reaches the dorsal head (vertex). The articulation that separates clypeus from labrum is not always visible in L1 larvae. A set of sensory organs provide cuticular markers for the lateral and dorsal sides of the Tribolium L1 larval head (Figs. 1D–E). In 62 wild-type cuticles, the entire set could be identified proving a high degree of constancy and reproducibility (exceptions are indicated below and are marked by open circles in Fig. 1). In the following, all bristles of one side of the head are described in an order that facilitates orientation–their names are quoted when mentioned for the first time.

At the posterior rim of the head at a dorso-lateral position, a row of four (in most cases) campaniform sensillae projects anteriorly ("bell row"). Anteriorly, the bell row is followed by the "row bristle". Dorsal to the latter, the posterior most of a triangle of three setae is found ("vertex triplet") composed of a "posterior vt", "anterior vt" and "ventral vt". Close to the posterior vt, the "posterior vertex bristle" is located. The "median vertex bristle" is found on an imaginary line between the two ventral vt of both sides. Dorsal to the anterior vt, the "anterior vertex bristle" is found. On an imaginary line from the ventral vt toward the base of the antenna, the "antenna basis bristle" is located. The bristles of the "labrum quartet" are located on an anterior extension of a line that runs through the "posterior vt" and the "anterior vt". The "gena triplet" is found ventral to the vertex triplet and ventro-anterior to the bell row. It marks the lateral portions of the head (gena). It is composed of "anterior gt", "posterior gt" and "dorsal gt" and in many cases encloses a campaniform sensillum (open circle). In most cases, the "gena bristle" is found between the anterior gt and

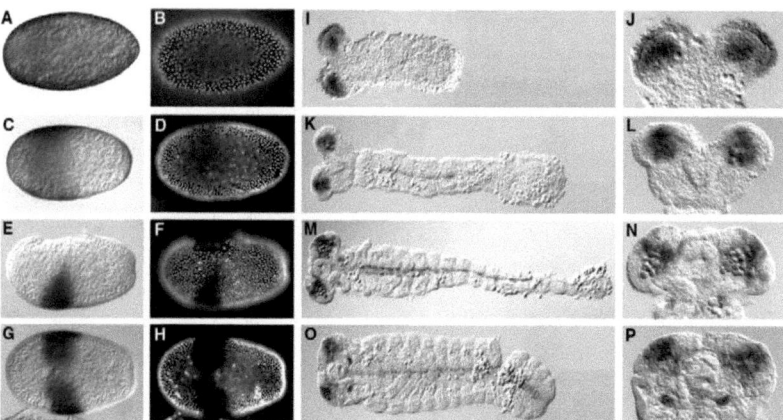

Fig. 2. Expression of *Tc-otd1*. (A, B) *Tc-otd1* is provided maternally. (C–H) The initially broad domain rapidly retracts from both poles and comes to lay at the anterior portion of the head. Panels E and F are lateral views, panels G and H are ventral views. (I, K, M, O) During germ band growth, the head domain remains rather stable while a midline domain arises de novo. (J, L, N, P) Close ups of the respective heads.

the maxilla. Three short setae line the maxillary bulge ("anterior, median and posterior maxilla escort") but depending on the orientation of the head they are sometimes difficult to identify on one side. The dorsal median region is poor in markers. The "dorsomedian bristles" are not visible on all wild-type cuticles. The larval eyes are located below the cuticle and we have not been able to detect a cuticular marker of the eye itself. However, in a lateral view, the eye appears in a field defined by four sensory organs: the ventral vt, the antenna basis bristle and the dorsal and anterior gt. In dorsal views, the eye appears below to the ventral vt.

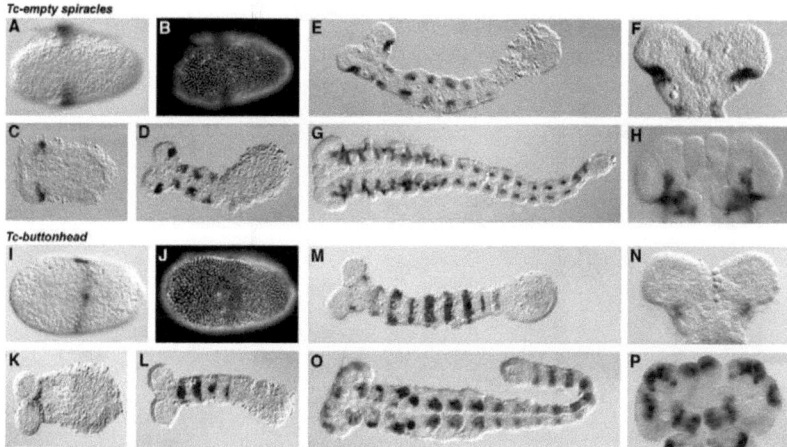

Fig. 3. Expression patterns of *Tc-ems* and *Tc-btd*. (A, B) *Tc-ems* starts to become expressed in the late blastoderm stage in a narrow stripe. (C) This stripe comes to lie in the anterior portion of the antennal segment anlagen. (D, E, G) During elongation, a segmentally reiterated pattern of lateral *Tc-ems* patches arises. (F, H) Close ups of the heads of embryos shown in panels E and G, respectively. (I, J) Also *Tc-btd* expression starts as a narrow stripe somewhat earlier than *Tc-ems*. (K) This stripe marks the future mandible. (L, M, O) During elongation, segmental stripes arise in anterior posterior sequence. (M, O) Midway through elongation, also the more anterior antennal and intercalary stripes appear and the appendages get *Tc-btd* positive. (N, P) Close ups of heads of the embryos shown in M and O, respectively. At late stages, *Tc-btd* becomes expressed in the labrum and several domains in the anterior head.

Identifying orthologs of orthodenticle, empty spiracles and buttonhead

Two *Tribolium* orthologs of the single *Drosophila otd* gene have been described (*Tc-otd1* and *Tc-otd2*) (Li et al., 1996). Expression of *Tc-otd2* starts only at the extended germ band stage when the head has already formed and is therefore unlikely to contribute significantly to early head patterning.

Using degenerated primers and scrutinizing the genomic sequence, we identified three SP factor genes. One being identical to the described *Tc-SP8* (Beermann et al., 2004). Phylogenetic analysis including all *Drosophila* and mouse SP factors reveals single *Tribolium* SP8 (Beermann et al., 2004) and *Drosophila* SP8 (originally termed *D-Sp1*; Wimmer et al., 1996; Schöck et al. 1999) orthologs to the mouse SP8/9 family that probably also includes the SP7 gene (Fig. S1D). *Dm*- and *Tc-SP1234* are likely orthologs to the mouse *Sp1/2/3/4* genes. The mouse SP5 gene is the closest related gene to the third *Tribolium* SP-factor (called Tc-Btd) but orthology is not unequivocal based on sequence analysis restricted to the zinc finger and *buttonhead* box. Supporting this assumption, however, 11 amino acids of the SP-box are identical between Tc-Btd and SP5 but 5 maximum with the other mouse SP-factors, respectively (not

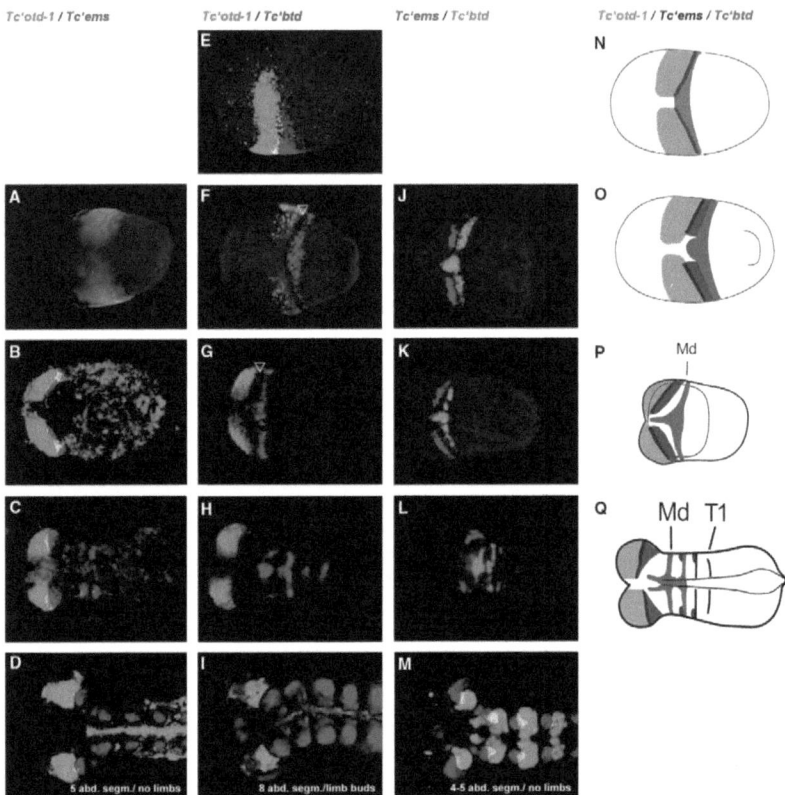

Fig. 4. Double in situ hybridization of *Tc-otd1*, *Tc-ems* and *Tc-btd*. False color representation of double stainings combining a conventional with a fluorescent staining. In all panels, anterior is to the left, all but panel E are ventral views. Comparable stages are shown in one row. Exceptions are panel J which is slightly older than panels A and F and D which is much younger than panels I and M. (A–D) *Tc-otd1* in green and *Tc-ems* in red. *Tc-ems* expression starts at a late blastoderm stage adjacent to *Tc-otd1* with a small overlap (A). This relative position including overlap remains constant throughout initial elongation (B, C). Later, *Tc-ems* expression fades at the lateral sides such that the overlap with *Tc-otd1* remains only in medial parts of its domain (D). (E–I) *Tc-otd1* in green and *Tc-btd* in red. (E) *Tc-btd* expression starts earlier than *Tc-ems* and only in its very first stages shows some overlap with *Tc-otd1*. (F) Slightly later, the stripe has become even narrower and the overlap with *Tc-otd1* is lost. In the opening clearance, *Tc-ems* will arise (compare open arrowhead in F with A and J). (G, H) The *Tc-btd* stripe comes to lay in the mandibular segment and the clearance to the *Tc-otd1* domain becomes broader (see open arrowhead). (I) At later elongation stages, *Tc-btd* stripes are found also in intercalary and antennal segments closing the gap to *Tc-otd1* again. (J–M) *Tc-ems* expression shown in red and *Tc-btd* in green. (J) In late blastoderm and early germ rudiment stages (shown in panel J), *Tc-btd* and *Tc-ems* expression is adjacent, probably without overlap. (K, L) In subsequent stages, the gap between these expression domains broadens. (M) Later, segmental expression is similar in all segments and overlap is observable in each segment. (N–Q) Schematic representation of our analysis with *Tc-otd1* shown in green, *Tc-btd* in red and *Tc-ems* in blue.

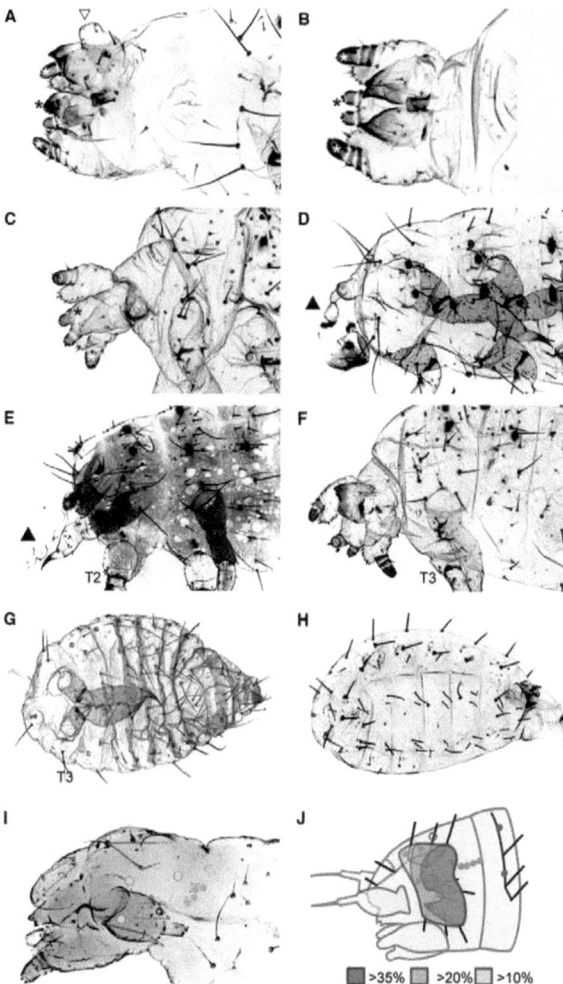

Fig. 5. *Tc-otd1* RNAi cuticle phenotype. The *Tc-otd1* phenotypic series ranges from deletions of single bristles to loss of entire head and thorax. The labium and the maxillae are marked by black and white stars, respectively. (A) Dorsolateral view of an intermediate phenotype that has lost the antennae and the entire dorsal and lateral setae but retains the labrum (open arrowhead). (B) Dorsal view of a phenotype where all head structures but the gnathal appendages are deleted. (C) Lateral view of a cuticle which only retains labium and maxillae. (D) Ventrolateral view on a cuticle the head of which is reduced to a cuticular tube (black arrowhead). (E, F) In stronger phenotypes, also the thoracic segments are affected. More anterior structures are reduced to a cuticular tube (E). In rare cases, some gnathal segments remain but thoracic segments are lost (F). (G, H) Strong phenotypes lack the entire head and parts (G) or the entire thorax (H). Some knockdown embryos loose any sign of segmentation and form cuticular sacs decorated with some bristles (not shown). (I, J) Weak phenotypes that retain antennae and labrum were scored for the head bristle pattern. (I) Labrum and distorted antennae are present. Missing setae are posterior vt, all three gena triplet setae and the posterior maxilla escort. Missing bristles are marked by red circles. (J) Schematic summary of the analysis. All bristles and setae that were lost in >35%, >20% and >10% are combined in colored fields, respectively (n=28). Because additional markers are lacking, the extension of these fields has been chosen to just comprise the missing bristles. A cuticle field extending from the lateral head to the dorsum in the middle of the head is most sensitive. Less affected are regions anterior to this field. Detailed results for the bristle analysis are found in Table S1.

shown). Even more equivocal is the association of Drosophila Btd with other SP-factors, maybe due to the high evolutionary rate of Dm-Btd (see long branch in Fig. S1D). We see only remnants of the SP-box in Dm-Btd (Fig. S1A). However, the genomic location of both Drosophila and Tribolium btd genes close to their SP8 paralogs (Schöck et al., 1999; The Tribolium Genome Sequencing Consortium, in press), respectively, and their similar expression patterns (see below) argue in favor of orthology. Therefore, we are convinced that Tc-btd is the true ortholog of Dm-btd (see Fig. S1A for additional findings).

Tc-ems was isolated in a library screen using the Drosophila sequence as a probe (Hausdorf, 1996). No additional paralog is found in the genomic sequence. In the Drosophila genome, however, an additional gene with a high similarity is present, Dm-E5 (CG 9930). Our phylogenetic analysis reveals Tc-Ems as the single ortholog to the paralogs Dm-E5 and Dm-Ems (Fig. S1C). The Drosophila gene pair probably arose by duplication after the separation of Dipteran and Coleopteran ancestors. The mouse paralogs emx1 and 2 have probably arisen by a recent independent duplication event (Williams and Holland, 2000). Interestingly, the length of Tc-Ems is more similar to the short mouse proteins rather than to the much longer Drosophila proteins. These Drosophila specific expansions have most likely occurred independently because they are located at different locations between box 1 and 2 (Dm-Ems) and at the C-terminus (Dm-E5). Dm-E5 is not transcribed in blastoderm stages but becomes expressed in a segmentally reiterated pattern similar to late Dm-ems expression from stage 10 onward (Williams and Holland, 2000) and is therefore not relevant for our discussion.

Different onset of expression of the head gap gene orthologs

Tc-otd1 expression has been described before (Li et al., 1996; Schröder, 2003). It differs from Dm-otd by its ubiquitous maternal contribution (Figs. 2A, B). During advancing blastoderm stages, this expression retracts from both poles clearing the anlagen of the extraembryonic tissues (anterior) and the posterior portion of the embryo (Figs. 2C–H) and also from ventral tissue (Fig. 2G). From this stage on, the head expressions of Dm-otd and Tc-otd1 are very similar. At later stages, activity along the midline and in the anterior portion of the mandibles arises (Figs. 2K–P).

Expression of Tc-ems starts at the late blastoderm stage, when extraembryonic tissue and germ rudiment become morphologically distinguishable (Figs. 3A, B). This stripe is narrow and sharp from the beginning and remains adjacent to the ocular Tc-wg stripe in germ bands without detectable overlap. Also the later antennal Tc-wg stripe touches Tc-ems expression without overlap (not shown). This locates the first Tc-ems domain to the anterior portion of the antennal and the posterior most portion of the ocular segment. The non-overlapping adjacent expression of wg and ems is also seen in the head ectoderm of Drosophila stage 10 embryos (Urbach and Technau, 2003a, b). During germ band growth, additional segmental patterns arise in the lateral portions of gnathal and trunk segments that appear similar to the Drosophila expression but are not further analyzed here (Figs. 3D, E, G).

Tc-btd expression starts in the late blastoderm stage as well but arises somewhat earlier than Tc-ems (Figs. 3I, J). In the differentiated blastoderm stage, it forms a narrow stripe that later comes to lie in the mandibular segment (Fig. 3K). Subsequently, segmental stripes arise (Figs. 3L, M, O). Rather late, expression appears also in the antennal segment (Fig. 3M) and the ocular region, the labrum and in the anterior head (Figs. 3O, P). Initially, the future mandibular Tc-btd stripe is separated from the ocular Tc-wg domain only by a few cells (not shown) but this gap extends significantly in subsequent stages.

By a series of double stainings, we asked whether Drosophila-like extensive overlap of expression patterns (Wimmer et al., 1997) is found in Tribolium (Fig. 4). Shortly before the extraembryonic tissue

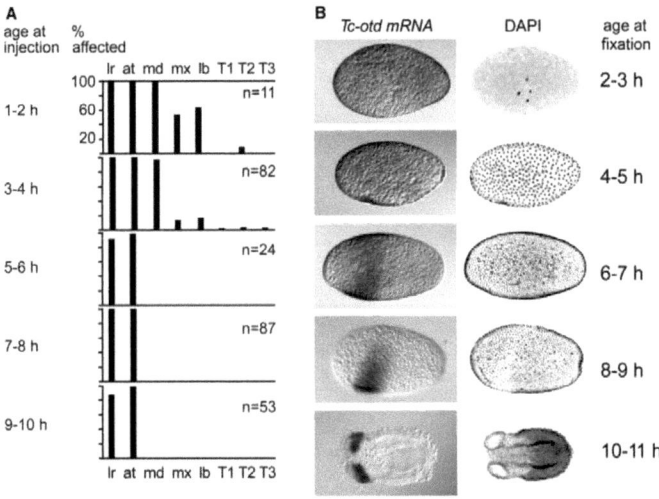

Fig. 6. Early regionalization defects depend on knock down of early Tc-otd1 function. (A) Embryos were injected at different time points after egg collection and the cuticle phenotype scored for the presence of the head and trunk appendages. Only injections within the first 4 h of development (at 32 °C) lead to defects of gnathal and thoracic segments. Later injections lead only to loss of antenna, eye and labrum. (B) Staining of similarly staged embryos for Tc-otd1 expression. Assuming a short delay between injection, mRNA degradation and protein decay, we find that the retraction of Tc-otd1 expression to the anterior head (6–7 h) correlates with the more restricted deletion of antenna, eye and labrum.

begins to become morphologically visible, the mandibular Tc-btd stripe arises. Initially, it overlaps with the still broad Tc-otd1 stripe (Fig. 4E) but shortly later, a gap arises between Tc-otd1 and Tc-btd where the Tc-ems expression becomes detectable (Figs. 4F, A, J; J is slightly older than A, F). Initially, the newly formed Tc-ems and the Tc-btd stripes are adjacent but non-overlapping. Subsequently, the Tc-btd expression domain becomes separated from Tc-ems expression (Fig. 4, compare J with K). This gap broadens with time (Fig. 4L) leaving space for posterior antennal and intercalary segment anlagen. At mid elongation, the antennal and intercalary Tc-btd domains arise (Figs. 4I, M; I is older than M and D) that fill parts of the gap and lead to a repetitive pattern that is similar from antenna to posterior trunk segments. From that stage on, Tc-btd and Tc-ems overlap to some extent. The Tc-ems and Tc-otd stripes show some small overlap throughout development (Figs. 4A–D).

In summary, the Tribolium orthologs are expressed in blastoderm stages in single stripes in the head anlagen but in contrast to Drosophila only Tc-otd1 transcripts are detected in early blastoderm stages in a dynamic pattern indicative for early regionalization. Tc-ems and Tc-btd initiate later in already distinct stripes that do not cover more than one segment primordium. Hence, their expression patterns are not in line with a role in early regionalization events.

Divergent aspects of Tc-otd1 function correlate with changes of expression pattern

To identify the function of these genes in head development, we knocked down their transcripts using both parental (pRNAi) and embryonic RNAi (eRNAi). To verify that Tc-otd2 does not contribute to head patterning, we knocked down its transcript via RNAi. Almost all embryos hatched and all showed the wild-type bristle pattern. The lack of an RNAi phenotype is not surprising regarding its late onset and restricted pattern of expression (Li et al., 1996). Therefore, we have excluded Tc-otd2 from further analysis.

The Tc-otd1 phenotype has been described to range from a gap like deletion of ocular and antennal segments up to the deletion of the entire head (Schröder, 2003). We find this phenotypic range (Figs. 5A–F) but also detect stronger phenotypes in pRNAi experiments (Figs. 5G, H) confirmed by eRNAi performed 2–3 h after egg laying. Strong phenotypes lack not only the head, but also parts of the thorax (Figs. 5G, H). In about 20% of the RNAi treated embryos, we find cuticles with severe disturbances where only few thoracic and abdominal segments are left or even non-segmented sack-like cuticles with some residual bristles (not shown). A large portion of the RNAi phenotypes (63%) display posterior segmentation defects. The dorsal

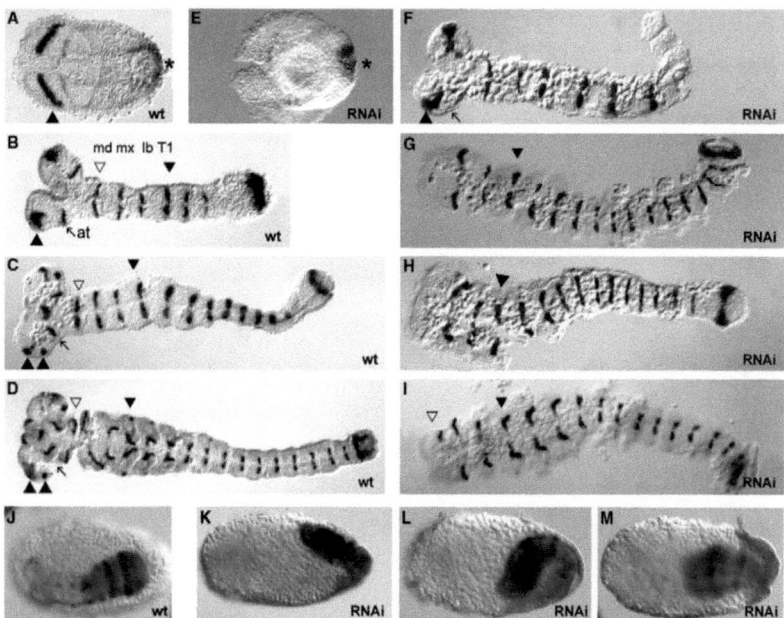

Fig. 7. Tc-wg in Tc-otd1 RNAi knockdown embryos reflect the cuticular defects. All embryos are shown with anterior to the left. The mandibular segment is marked with an open arrowhead, the first thoracic segment with a black arrowhead. The ocular Tc-wg domain is marked with a black arrowhead while the antenna is indicated with an arrow. (A–D) Wild-type embryos at different stages of elongation. (E) Young germ band showing intact growth zone expression (star) but absent ocular Tc-wg domain. The head tissue appears to be less stable and deranged. (F) Weak phenotype that retains the ocular (arrowhead) but has lost the antennal Tc-wg domain (arrow). (G–I) Embryos in the extended germ band stage displaying different grades of anterior deletions of head and gnathal segments. (J–M) Tc-hairy staining in wild-type (J) and RNAi embryos (K–M) in approximately similar stages. The loss of head tissue leads to a drastic immersion of the growth zone into the yolk leading to a situation where the head remnants are at the posterior of the blastoderm while the posterior growth zone is directed toward anterior.

cuticle of thorax and anterior abdomen appear to be especially sensitive to Tc-otd1 knockdown. In some cases, also more posterior segments of the abdomen are affected or deleted (not shown). In Tribolium, some strong segmentation or dorso-ventral phenotypes do not secrete cuticle. In order to assess in how far we miss potentially strong phenotypes in our cuticle analyses, we counted the portion of cuticles and empty egg shells after pRNAi. We find that both in wild-type and RNAi egg collections 30-38% of the eggs do not develop a cuticle which is a slightly elevated number. As we focus on head development in this work, we have not followed these defects further.

To map the head defects, we analyzed 24 cuticles that showed remnants of the head. Most of them (46%) lack all head structures but maxilla and labium (Fig. 5C). The maxilla appears more stable than the labium as it is often the only head segment present (29.1% versus 8.3%). Our data do not reveal if this is due to more stable patterning of the maxillary segment or to homeotic transformations. In 16.6%, the entire gnathum (mandible, maxilla and labium) is present while the antenna is missing (Fig. 5B). The non-segmental labrum remains present in 21% of the cuticles (Fig. 5A) either with or without antenna. To track down the weakest effects of Tc-otd1 RNAi, we scrutinized 28 cuticles that displayed all head structures (labrum and all head appendages) for the bristle pattern described above (Table S1 and Figs. 5I, J). We find that the most sensitive region is marked by posterior and ventral vt, dorsal and posterior gt, the bell row bristle and the posterior and median maxilla escort bristles. Even in weak phenotypes, the eyes are usually absent. In stronger phenotypes, the defects extend toward anterior until they lead to loss of antenna, labrum and mandibles (see above). This type of deletion then usually includes loss of all vertex and gena bristles and the bell row.

Interestingly, the weaker Tc-otd1 RNAi phenotypes are similar to the Drosophila mutants where ocular and antennal structures are affected.

We wondered whether the maternal expression that is found in Tribolium but not Drosophila might be responsible for the strong phenotypes. To knock down the late Tc-Otd1 function without affecting the early maternal contribution, we injected embryos at different time points after egg laying and determined the portion of missing labrum, head and thoracic appendages in cuticle preparations. Indeed, the early injections (1-4 h after egg laying) lead to deletions of anterior segments up to the third thoracic segment while injections after 5 h elicit the weaker Drosophila like deletions (Fig. 6A). We were not able to stain injected blastoderm stages to determine the exact time point of transcript degradation upon RNAi and we do not know the dynamics of Tc-Otd1 protein turnover. Assuming a lag of 1-2 h, we estimate the more restricted requirement for Tc-otd1 to start at 5-6 h. This analysis suggests that the Tc-otd1 expression after retraction from both poles (Fig. 6B, 6-7 h) correlates with the more restricted RNAi phenotype (Fig. 6A, 5-6 h). From that time on, Tc-otd1 expression is similar to Drosophila in location and extent and also the deletion domain of mutants/ RNAi phenotypes covers approximately the same region (see Fig. 10).

To get more insight into the embryonic origin of the phenotype, we performed Tc-wg stainings in knockdown embryos. Essentially, the results reflect the defects seen in cuticles. In young embryos undergoing involution, the growth zone appears to be correctly specified and the posterior Tc-wg stripe is always present (black star in Figs. 7A, E). However, the anterior tissue is not properly formed and the ocular Tc-wg stripe is often missing (Fig. 7E). The loss of the entire head is reflected in deletions of the respective Tc-wg stripes in fully extended germ bands (Figs. 7G-I). We also find correlates for weaker deletions

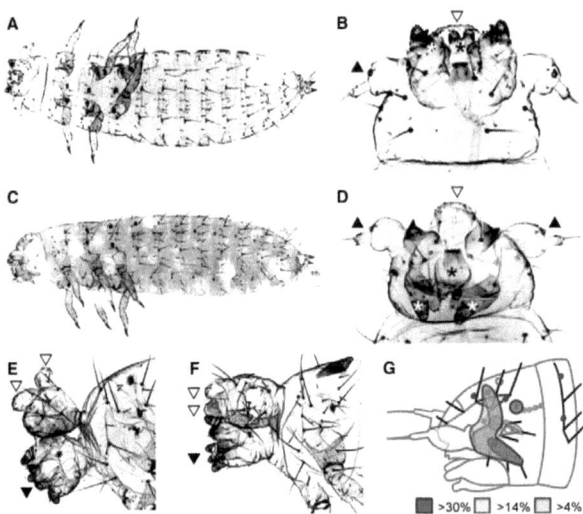

Fig. 8. Tc-ems RNAi leads to defects in the antennal segment. Knock down of Tc-ems function leads to mild phenotypes where the antennae are twisted toward posterior and are not properly formed at their basis. (A, C) Ventral and ventrolateral view of cuticles the head of which are shown in panels B and D. Labrum (open arrowhead), labium (black star) and maxillae (white stars) appear unaffected. (E, F) In rare cases, more tissue is deleted leading to embryos that have lost the antennae and remain with two separate lobes. A ventral lobe formed by the gnathal appendages (black arrowhead) and a dorsal lobe including dorsal cuticle and the labrum that is sometimes split (open arrowheads) and appears to contain the vertex triplet bristles (purple triangle). Because these defects occur rarely (3 cuticles of 23) and are not linked to the twisted antennae phenotype by intermediates, we assume that these defects are due to secondary morphogenesis defects. (G) The bristle pattern shows a deletion pattern partially overlapping with Tc-otd1 (see Fig. 5). In addition, we find unusually frequent duplications of bristles in an additional more posterior region (>10% shaded in grey). Detailed results for the bristle analysis are found in Table S1.

as for instance the loss of the antennal stripe (Fig. 7F) or the antennal together with the mandibular stripes (not shown). Curiously, the deletion of anterior tissue apparently leads to an atypical twisting and immersion of the growth zone into the yolk (Fig. 7, compare K–M with J). In its extreme form, the twisting leads to an inversion of the embryonic axis relative to the egg axis (Fig. 7M). Curiously, a superficially similar rearrangement of the embryo within the egg is also observed in Tc-zen1 RNAi (van der Zee et al., 2005). However, we do not see loss of serosa tissue in Tc-otd1 RNAi (as judged by the wider spaced nuclei) but we observe rupturing of embryonic tissues anterior to the growth zone (not shown). An exact analysis of these early events is beyond the scope of this paper but apparently different kinds of changes in blastodermal tissue composition can lead to similar rearrangements of the growth zone within the egg.

Tc-ems function is restricted to the antennal segment

pRNAi and eRNAi knock down of Tc-ems leads to surprisingly mild cuticular phenotypes. The defect is marked by antennae with poorly formed basal segments that are posteriorly bent (Fig. 8) and the eyes are usually absent (not shown). The median and posterior bristles of the maxilla escort are most sensitive to Tc-ems RNAi. In stronger phenotypes, the defects extend dorsally including the eye, the ventral vt and median vertex bristle (Table S1 and Fig. 8F). In rare cases (14.3%), the antennae are lost together with a median portion of the head leading to a situation where a ventral lobe consisting of gnathal segments is separated from a dorsal lobe often consisting of a distally split labrum and other dorsal structures (Figs. 8E, F). The respective bristle pattern is strongly deranged and thus hard to be analyzed. The

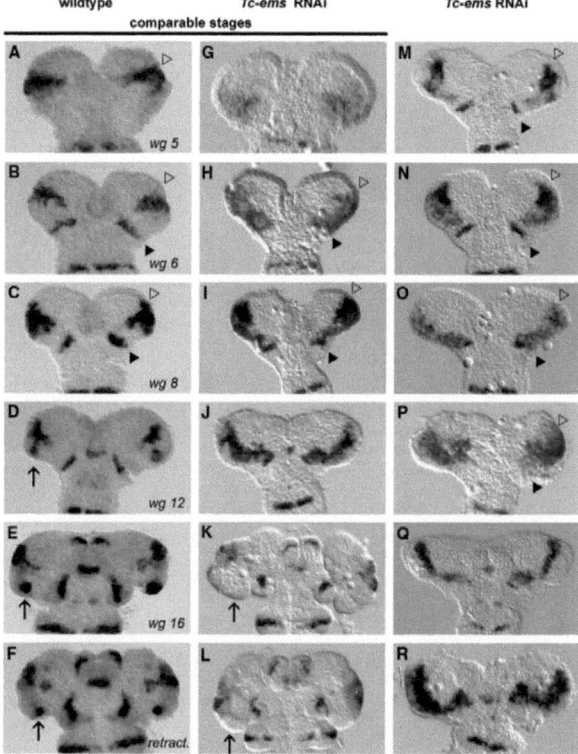

Fig. 9. Tc-ems determines the posterior border of the ocular Tc-wg domain. Heads of wild-type (A–F) and Tc-ems RNAi embryos (G–R). The age of embryos in panels G–L is comparable to the wild types panels A–F. The number of trunk Tc-wg stripes is given. Panels M–P are approximately the same stage but with different phenotypic strengths. The age compares to panels B and C. The age of the embryos in panels Q and R corresponds to panel D. The anterior border of the ocular Tc-wg domain is marked by an open arrowhead, the antennal stripe by a black arrowhead. (G–J) In early stages, the ocular Tc-wg domain is shifted posteriorly and is fused to the antennal Tc-wg domain. The posterior border of the antennal segment and the anterior border of the ocular segment appear to be unaffected (compare to panels A–D). (K, L) At later stages, the antennal stripe separates and becomes part of a shortened antenna. In wild type, the Tc-wg domain splits in two domains (E, F), the anterior of which is present in Tc-ems RNAi embryos but the posterior one is absent (compare arrows in panels E and F with panels K and L). (M, N) In weak phenotypes, the antennal Tc-wg stripe is unaffected and also the anterior border of the ocular Tc-wg domain appears normal (compare open arrowheads of panels B and C). (O–P) In stronger knockdowns, the ocular and antennal Tc-wg domain fuse and form one domain. (Q, R) This fused domain becomes a U shaped appearance in slightly older stages (comparable to panel D).

vertex setae, however, appear to be present (see purple triangle in Fig. 8E).

To get further insight into the embryonic development of the defect, we analyzed Tc-wg patterns in knockdown embryos (Fig. 9). We find that in weak phenotypes the posterior border of the ocular Tc-wg domain extends toward posterior while the antennal domain remains normal (Figs. 9I, M, N). In stronger phenotypes, these two domains fuse completely (Figs. 9H, O–R). Neither the anterior border of the ocular nor the posterior border of the antennal stripes appear to be affected although the highly dynamic ocular Tc-wg pattern makes it difficult to assess this unequivocally (compare Figs. 9A–F with G–L). Later, ocular and antennal domains appear to separate again (Figs. 9Q, R compare to D), giving rise to a distinct but shortened antennal Tc-wg stripe (Figs. 9K, L). Also the antenna itself is reduced and appears rounded rather then elongated (compare Figs. 9 K, L with E, F). In fully extended germ bands, the ocular Tc-wg domain splits into several domains, the posterior most of which probably marks the posterior boundary of the developing larval eye (Dong and Friedrich, 2005). This domain is clearly absent in Tc-ems RNAi embryos (black arrow in Figs. 9K, L and E, F) correlating with the absence of eyes in Tc-ems phenotypes. The anterior domain, in contrast, appears rather normal. Despite the segmentally iterated expression of Tc-ems, we do not find alterations of Tc-wg expression in the more posterior segments.

Tc-btd is not required for head development

Strikingly, we were not able to detect an RNAi phenotype for Tc-btd. To confirm this negative result, we stained knockdown embryos with a Tc-btd probe (including Tc-caudal as positive control in the same color reaction) and Tc-wg. Indeed, Tc-btd was knocked down below the limit of detection of whole mount in situ hybridization while both control staining and the Tc-wg pattern appeared normal (not shown). By counting laid eggs per female in comparison to a control (Tc-ems pRNAi), we confirmed that the injected beetles are not sterile and that a normal portion of embryos develops a cuticle (not shown). This suggests that the lack of cuticular phenotypes is not due to sterility of the injected animals or early embryonic death before the secretion of cuticle. In summary, we consider it likely that Tc-btd is not required for head development although we are aware of the inherent difficulty of proving the absence of gene function by RNAi (see materials and methods).

Discussion

Head development is not well understood in Drosophila because involution and derived morphology of the larval head have hampered the interpretation of mutant phenotypes. We have established the red flour beetle Tribolium castaneum as a model system for insect larval head development and asked how far the functions of the Tribolium orthologs of the known Drosophila head gap genes otd, ems and btd are conserved.

Tc-btd and Tc-ems are not head gap genes

Our most unexpected finding is that the Tribolium ortholog of btd is not required for head epidermis patterning. We did not see cuticular phenotypes although we were able to knock down Tc-btd below the detection limit of in situ hybridization. However, it is impossible to unequivocally confirm negative RNAi data as we cannot formally exclude that some Tc-Btd protein might have formed and been able to fully rescue a potential phenotype. Therefore, our interpretation remains arguable to some degree. However, the expression pattern supports our interpretation. In contrast to Drosophila, Tc-btd expression starts only at late blastoderm stages in a narrow stripe restricted to the future mandibular segment (Fig. 10). Only much later also antennal and intercalary stripes arise. This is not suggestive for an extensive early function in anterior head patterning. Interestingly, overexpression of Dm-btd in Drosophila does not lead to segmentation defects. This suggests that even in the fly the borders of btd expression are not instructive for metamerization (Wimmer et al., 1997). Our sequence analysis of the zinc finger and the buttonhead box has assigned clear orthology to Tc-SP8 and Tc-SP1234 and has revealed that the third factor (Tc-Btd) is the closest homolog of Dm-Btd in the Tribolium genome. However, these latter two genes do not cluster on one branch in the tree as would be expected of orthologs (Fig. S1D). The long branches of both genes indicate that they have undergone accelerated sequence evolution that has blurred their origin. Further support for our orthology statement comes from domain architecture, genomic location and late aspects of expression that are identical for both genes. Our notion that btd could be an SP5 ortholog remains to be tested by the inclusion of SP5/btd orthologs from a broader arthropod sampling. But we note similarities of mouse SP5 and Tc-btd in that both display dynamic embryonic expression patterns but lack overt phenotypes (Harrison et al., 2000; Treichel et al., 2001).

We find that Tc-ems is not a head gap gene because it is required only for the posterior portion of the ocular and the anterior portion of the antennal segment by positioning the posterior border of the ocular Tc-wg domain. Of the antenna itself, only the anterior most tissue is affected. This is likely to result in a one-sided disturbance of the outgrowth which in turn might lead to the bent antenna phenotype that we observe. The genesis of the strong phenotype with the dorso-ventrally split head remains obscure: we assume, however, that this is a secondary effect of loss of ocular and antennal tissue: The antennal tissue is located at the anterior end of the head that connects the upper head capsule with the gnathum below. Missing tissue there could lead to a loss of contact of the upper and lower part while both tissues continue development. Finally, both parts might separately close the "holes" that arise by the separation by formation of ectopic cuticle. Unfortunately, the morphogenetic movements that transform the head anlagen of the germ band to the final head are complicated and completely unknown so far—a closer description of these

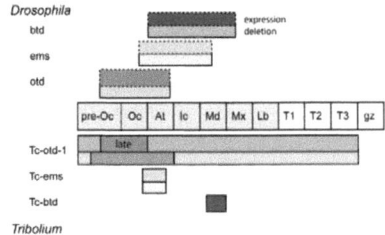

Fig. 10. Expression and deletion domains of Drosophila and Tribolium otd, ems and btd at early embryonic stages. Expression and deletion domains are aligned relative to the fate map of the embryo shown in the middle (pre-Oc: preocular region including labrum, segments: Oc: ocular, At: antennal, Ic: intercalary, Md: mandibular, mx: maxillary, Lb: labial, T: thoracic, gz: growth zone). The upper bars with more intense color indicate expression, the lighter bar below indicates the deletion domain in mutants or RNAi knockdowns. Expression of the Drosophila head gap genes is depicted as described for stage 5 embryos. They display overlapping expression and deletion patterns. However, the alignment of the expression domains to segment primordia has not been exactly determined and is based on the assumption that expression and deletion domains coincide (Wimmer et al., 1997) (hatched outline). The Tribolium orthologs are expressed in the same anterior posterior order but without significant overlap. The deletion domain of Tc-ems is restricted to the posterior part of the ocular and the anterior portion of the antennal segment while Tc-btd is not required for head development. Tc-otd1 has an early broad expression domain that retracts to tissues including, and anterior to, the ocular segment (dark green). We find an early regionalization function that affects the entire blastoderm fate map and a later and more restricted head patterning function. The transition is apparently gradual. See text for more details.

processes is badly needed. In any case, the later and more restricted expression in the *Tribolium* blastoderm correlates with a narrower deletion domain when compared to *Drosophila* where *ems* is required for patterning the antennal, intercalary and parts of the ocular segment (Fig. 10).

Two phase function of Tc-otd1

Tc-otd1 is required for patterning large parts of the anterior embryo (Schröder, 2003). We find phenotypes that exceed the effects described before. They lack the entire head and thorax, show additional abdominal segmentation defects or in rare cases do not form segments at all. Interestingly, the severe deletions in embryonic injections only occur, when *Tc-otd1* dsRNA is injected within the first 4 h of development (at 32 °C) which correlates with the phase of broad and dynamic blastodermal expression of *Tc-otd1* (Fig. 6). We therefore suggest a two phase function of *Tc-otd1*: an "early regionalization function" and a later and more restricted "head patterning function". In the first phase, *Tc-otd1* could be required for partitioning the germ rudiment into non-growth zone versus growth zone tissue. In line with this model, we find phenotypes that lack all segments specified in the blastoderm (head and thorax) while the growth zone dependent abdominal segments are formed. The additional posterior patterning defects are unlikely to be direct effects, because *Tc-otd1* is not active at the posterior pole or in abdominal tissues at any time (apart from the midline expression). However, a massive loss of tissue in the anterior germ rudiment could affect growth zone integrity which secondarily could lead to posterior segmentation defects. Indeed we find that the growth zone is unnaturally turned and immersed into the yolk mass in knockdown embryos (Figs. 7K, L). A study of cell behavior and marker genes for the growth zone in knockdown embryos is required to test this hypothesis. The head patterning function of *Tc-otd1* is restricted to the ocular and antennal regions and hence is similar to the *Drosophila* function. One clear difference to *Drosophila* is the role of *Tc-otd1* for labrum development (Cohen and Jürgens, 1990; Finkelstein et al., 1990; Schröder, 2003). As the labrum anlagen are located in a tissue adjacent to *Tc-otd1* expression, the requirement could be indirect.

What is the ancestral state of *orthodenticle* expression? The late *otd* expression appears to be widely conserved among arthropods as for instance in *Parhyale hawaiiensis* (Crustacea) (Browne et al., 2006) and several chelicerates: the oribatid mite *Archegozetes longisetosus* (Telford and Thomas, 1998), the spider *Tegenaria saeva* and the scorpion *Euscorpius flavicaudis* (Simonnet et al., 2006). Also contribution to early regionalization is found throughout arthropods: Like in *Tribolium*, the hymenopteran *Nasonia vitripennis* ortholog is maternally expressed but in contrast to the beetle, its mRNA is localized to the anterior and also the posterior pole and plays important morphogenetic functions at both (Lynch et al., 2006). In the spider *Achaearanea tepidariorum*, *otd* is among the first genes to respond to the initial symmetry breaking event (Akiyama-Oda and Oda, 2003). Obviously, *otd* orthologs play a crucial role in early anterior patterning events in several arthropods but they perform in quite different ways. In contrast, *otd* expression in the scorpion *E. flavicaudis* starts at the 6 segment germ band stage in an already distinct anterior stripe arguing against a role at an earlier stage (Simonnet et al., 2006). Therefore, a requirement of *otd* for early anterior patterning is likely to be ancestral at least for holometabolous insects.

Different expression patterns correlate with different functions

In all three cases, we see a correlation of change in function with different expression patterns: The earlier and broader expression domains of *ems* and *btd* in *Drosophila* and *Tc-otd1* in *Tribolium* correlate with their requirement for patterning broader regions. This adds to the current view that cis-regulatory changes are crucial for the evolution of gene function (McGregor et al., 2007; Wray, 2007) and makes these genes interesting models to pin down the respective regulatory changes. Interestingly, it is the early aspect of expression that differs most between *Tribolium* and *Drosophila* while the late patterns appear conserved (Dalton et al., 1989; Finkelstein et al., 1990; Wimmer et al., 1993, 1996). There appears to be less constraint for evolutionary change of early patterning in insects. *orthodenticle* orthologs for instance have been found to be expressed zygotically (Cohen and Jürgens, 1990) or maternally, with either a ubiquitous (Li et al., 1996) or localized mRNA distribution (M.-F. Schetelig, B.G.M. Schmid, E.A.Wimmer, unpublished). Even mRNA localization to both poles has been found (Lynch et al., 2006) and a similar variety is seen for *giant* orthologs (Brent et al., 2007; Bucher and Klingler, 2004; Mohler et al., 1989). The lack of *bicoid* in most insects further supports the notion of highly evolvable early patterning in insects (Brown et al., 2001; Schröder, 2003).

Comparing gene function across bilaterian animals

The identification of genes that are involved in both vertebrate, annelid and insect anterior patterning has led to the suggestion of highly conserved mechanisms (Acampora et al., 1998; Denes et al., 2007; Reichert and Simeone, 1999; Simeone et al., 1992; Treichel et al., 2003; Wimmer et al., 1993). At first glance, the significance of such cross phylum comparisons is questioned by the variability that we detect even within holometabolous insects. However, we also show that this variability is mainly restricted to early patterning events while the late aspects of for instance *orthodenticle* expression are conserved between *Tribolium*, *P. hawaiiensis* (Crustacea) (Browne et al., 2006) and several chelicerates: the oribatid mite *A. longisetosus* (Telford and Thomas, 1998), the spider *T. saeva* and the scorpion *E. flavicaudis* (Simonnet et al., 2006). In the case of *ems*, the late expression pattern of insects is similar to the one in the spider *T. saeva* and the scorpion *E. flavicaudis* (Simonnet et al., 2006) and can therefore be regarded as the ancestral state. The comparisons between brain phenotypes in vertebrates and insects upon *ems* depletion, however, are currently based on *Drosophila* mutants that interfere with both early head gap gene patterning and later function in the brain (Hirth et al., 1995). Hence, these authors have potentially observed a composite phenotype where the early and broad requirement of *ems* in the ectoderm (that still comprises the neuroectoderm at that stage) may have produced a larger deletion of the brain than the later more restricted activity in the brain may do on its own. More specifically, the *Tribolium* phenotype that is restricted to the posterior portion of the ocular and anterior portion of the antennal segments indicates that only part of the deutocerebrum may actually depend on *ems* function rather than the entire deuto- and tritocerebrum (Hirth et al., 1995; Reichert and Simeone, 1999). We therefore argue that cross phylum comparisons should not be based on the extremely variable early blastodermal function and expression. In contrast, later aspects may tend to be more conserved because patterning then converges to the specification of organ primordia or even specific cell types with respective molecular fingerprints (Arendt, 2005; Tessmar-Raible et al., 2007) rather than the broad subdivision of embryonic fields.

This work shows that the head gap gene paradigm of *Drosophila* is not valid for other arthropods. Because of its insect typical mode of embryonic head development, *T. castaneum* is becoming the primary model system for head development. Its amenability to reverse genetics allows us to identify the crucial genes from an extensive candidate gene list (e.g. comprising genes involved in vertebrate neural plate patterning and/or genes expressed in relevant stages and tissues in *Drosophila*). In addition, forward genetics by the ongoing GEKU insertional mutagenesis screen (Göttingen, Erlangen, Kansas State University, USDA) will reveal novel players by a hypothesis independent approach. Finally, a detailed understanding of *Tc-ems* and Tc-otd1 function and their potential interaction with the dorsoventral patterning system with its known effects on head development (van der Zee et al., 2006) is needed.

Acknowledgments

We thank Bernhard Hausdorf and Reinhard Schröder for providing the Tc-ems clone and Jonas Schwirtz for help with the Tc-wg wild-type staining. This work was supported by the DFG grants BU1443-2-2 (G. B.) and Wi1797/2-2 (E.A.W.).

Appendix A. Supplementary data

Supplementary data associated with this article can be found, in the online version, at doi:10.1016/j.ydbio.2008.03.005.

References

Acampora, D., Avantaggiato, V., Tuorto, F., Barone, P., Reichert, H., Finkelstein, R., Simeone, A., 1998. Murine Otx1 and Drosophila otd genes share conserved genetic functions required in invertebrate and vertebrate brain development. Development 125, 1691–1702.
Akiyama-Oda, Y., Oda, H., 2003. Early patterning of the spider embryo: a cluster of mesenchymal cells at the cumulus produces Dpp signals received by germ disc epithelial cells. Development 130, 1735–1747.
Arendt, D., 2005. Genes and homology in nervous system evolution: comparing gene functions, expression patterns, and cell type molecular fingerprints. Theory Biosci. 124, 185–197.
Beermann, A., Aranda, M., Schroder, R., 2004. The Sp8 zinc-finger transcription factor is involved in allometric growth of the limbs in the beetle Tribolium castaneum. Development 131, 733–742.
Brent, A.E., Yucel, G., Small, S., Desplan, C., 2007. Permissive and instructive anterior patterning rely on mRNA localization in the wasp embryo. Science 315, 1841–1843.
Brown, S.J., Mahaffey, J.P., Lorenzen, M.D., Denell, R.E., Mahaffey, J.W., 1999. Using RNAi to investigate orthologous homeotic gene function during development of distantly related insects. Evolut. Develop. 1, 11–15.
Brown, S., Fellers, J., Shippy, T., Denell, R., Stauber, M., Schmidt-Ott, U., 2001. A strategy for mapping bicoid on the phylogenetic tree. Curr. Biol. 11, R43–R44.
Browne, W.E., Schmid, B.G., Wimmer, E.A., Martindale, M.Q., 2006. Expression of otd orthologs in the amphipod crustacean, Parhyale hawaiensis. Dev. Genes Evol. 216, 581–595.
Bucher, G., Klingler, M., 2004. Divergent segmentation mechanism in the short germ insect Tribolium revealed by giant expression and function. Development 131, 1729–1740.
Bucher, G., Wimmer, E.A., 2005. Beetle a-head. B.I.F. Futura 20, 164–169.
Bucher, G., Scholten, J., Klingler, M., 2002. Parental RNAi in Tribolium Coleoptera. Curr. Biol. 12, R85–R86.
Budd, G.E., 2002. A palaeontological solution to the arthropod head problem. Nature 417, 271–275.
Cohen, S.M., Jürgens, G., 1990. Mediation of Drosophila head development by gap-like segmentation genes. Nature 346, 482–485.
Cohen, S., Jürgens, G., 1991. Drosophila headlines. Trends Genet. 7, 267–272.
Dalton, D., Chadwick, R., McGinnis, W., 1989. Expression and embryonic function of empty spiracles: a Drosophila homeo box gene with two patterning functions on the anterior-posterior axis of the embryo. Genes Dev. 3, 1940–1956.
Denes, A.S., Jekely, G., Steinmetz, P.R., Raible, F., Snyman, H., Prud'homme, B., Ferrier, D.E., Balavoine, G., Arendt, D., 2007. Molecular architecture of annelid nerve cord supports common origin of nervous system centralization in bilateria. Cell 129, 277–288.
Diederich, R.J., Pattatucci, A.M., Kaufman, T.C., 1991. Developmental and evolutionary implications of labial, deformed and engrailed expression in the Drosophila head. Development 113, 273–281.
Dong, Y., Friedrich, M., 2005. Comparative analysis of Wingless patterning in the embryonic grasshopper eye. Dev. Genes Evol. 215, 177–197.
Finkelstein, R., Perrimon, N., 1990. The orthodenticle gene is regulated by bicoid and torso and specifies Drosophila head development. Nature 346, 485–488.
Finkelstein, R., Smouse, D., Capaci, T.M., Spradling, A.C., Perrimon, N., 1990. The orthodenticle gene encodes a novel homeo domain protein involved in the development of the Drosophila nervous system and ocellar visual structures. Genes Dev. 4, 1516–1527.
Gallitano-Mendel, A., Finkelstein, R., 1998. Ectopic orthodenticle expression alters segment polarity gene expression but not head segment identity in the Drosophila embryo. Dev. Biol. 199, 125–137.
Gao, Q., Finkelstein, R., 1998. Targeting gene expression to the head: the Drosophila orthodenticle gene is a direct target of the Bicoid morphogen. Development 125, 4185–4193.
Gao, Q., Wang, Y., Finkelstein, R., 1996. Orthodenticle regulation during embryonic head development in Drosophila. Mech. Dev. 56, 3–15.
Griesel, G., Treichel, D., Collombat, P., Krull, J., Zembrzycki, A., van den Akker, W.M., Gruss, P., Simeone, A., Mansouri, A., 2006. Sp8 controls the anteroposterior patterning at the midbrain–hindbrain border. Development 133, 1779–1787.

Grossniklaus, U., Cadigan, K.M., Gehring, W.J., 1994. Three maternal coordinate systems cooperate in the patterning of the Drosophila head. Development 120, 3155–3171.
Haas, M.S., Brown, S.J., Beeman, R.W., 2001. Pondering the procephalon: the segmental origin of the labrum. Dev. Genes Evol. 211, 89–95.
Harrison, S.M., Houzelstein, D., Dunwoodie, S.L., Beddington, R.S., 2000. Sp5, a new member of the Sp1 family, is dynamically expressed during development and genetically interacts with Brachyury. Dev. Biol. 227, 358–372.
Hartmann, B., Hirth, F., Walldorf, U., Reichert, H., 2000. Expression, regulation and function of the homeobox gene empty spiracles in brain and ventral nerve cord development of Drosophila. Mech. Dev. 90, 143–153.
Hausdorf, B., 1996. In: Zoologisches Institut (Ed.), Charakterisierung von Entwicklungsgenen in Tribolium castaneum. Ludwig Maximilians Universität, München.
Hirth, F., Therianos, S., Loop, T., Gehring, W.J., Reichert, H., Furukubo-Tokunaga, K., 1995. Developmental defects in brain segmentation caused by mutations of the homeobox genes orthodenticle and empty spiracles in Drosophila. Neuron 15, 769–778.
Jurgens, G., Lehmann, R., Schardin, M., Nusslein-Volhard, C., 1986. Segmental organisation of the head in the embryo of Drosophila melanogaster. Roux's Arch. Dev. Biol. 195, 359–377.
Leuzinger, S., Hirth, F., Gerlich, D., Acampora, D., Simeone, A., Gehring, W.J., Finkelstein, R., Furukubo-Tokunaga, K., Reichert, H., 1998. Equivalence of the fly orthodenticle gene and the human OTX genes in embryonic brain development of Drosophila. Development 125, 1703–1710.
Li, Y., Brown, S.J., Hausdorf, B., Tautz, D., Denell, R.E., Finkelstein, R., 1996. Two orthodenticle-related genes in the short germ beetle Tribolium castaneum. Dev. Genes Evol. 206, 35–45.
Lichtneckert, R., Reichert, H., 2005. Insights into the urbilaterian brain: conserved genetic patterning mechanisms in insect and vertebrate brain development. Heredity 94, 465–477.
Lunardi, A., Vignali, R., 2006. Xenopus Xotx2 and Drosophila otd share similar activities in anterior patterning of the frog embryo. Dev. Genes Evol. 216, 511–521.
Lynch, J.A., Brent, A.E., Leaf, D.S., Pultz, M.A., Desplan, C., 2006. Localized maternal orthodenticle patterns anterior and posterior in the long germ wasp Nasonia. Nature 439, 728–732.
McGinnis, W., Krumlauf, R., 1992. Homeobox genes and axial patterning. Cell 68, 283–302.
McGregor, A.P., Orgogozo, V., Delon, I., Zanet, J., Srinivasan, D.G., Payre, F., Stern, D.L., 2007. Morphological evolution through multiple cis-regulatory mutations at a single gene. Nature 448, 587–590.
Mohler, J., Eldon, E.D., Pirrotta, V., 1989. A novel spatial transcription pattern associated with the segmentation gene, giant, of Drosophila. EMBO J. 8, 1539–1548.
Nassif, C., Daniel, A., Lengyel, J.A., Hartenstein, V., 1998. The role of morphogenetic cell death during Drosophila embryonic head development. Dev. Biol. 197, 170–186.
Reichert, H., Simeone, A., 1999. Conserved usage of gap and homeotic genes in patterning the CNS. Curr. Opin. Neurobiol. 9, 589–595.
Reichert, H., Simeone, A., 2001. Developmental genetic evidence for a monophyletic origin of the bilaterian brain. Philos. Trans. R. Soc. Lond., B Biol. Sci. 356, 1533–1544.
Rempel, G.J., 1975. The evolution of the insect head: the endless dispute. Quaest. Entomol. 11, 7–25.
Schmidt, H.A., Strimmer, K., Vingron, M., von Haeseler, A., 2002. TREE-PUZZLE: maximum likelihood phylogenetic analysis using quartets and parallel computing. Bioinformatics 18, 502–504.
Schmidt-Ott, U., Technau, G.M., 1992. Expression of en and wg in the embryonic head and trunk of Drosophila indicates a refolded band of seven segment remnants. Development 116, 111–125.
Schöck, F., Purnell, B.A., Wimmer, E.A., Jäckle, H., 1999. Common and diverged functions of the Drosophila gene pair buttonhead and D-Sp1. Mech. Dev. 89, 125–132.
Schöck, F., Reischl, J., Wimmer, E., Taubert, H., Purnell, B.A., Jäckle, H., 2000. Phenotypic suppression of empty spiracles is prevented by buttonhead. Nature 405, 351–354.
Scholtz, G., Edgecombe, G.D., 2006. The evolution of arthropod heads: reconciling morphological, developmental and palaeontological evidence. Dev. Genes Evol. 216, 395–415.
Schröder, R., 2003. The genes orthodenticle and hunchback substitute for bicoid in the beetle Tribolium. Nature 422, 621–625.
Simeone, A., Acampora, D., Gulisano, M., Stornaiuolo, A., Boncinelli, E., 1992. Nested expression domains of four homeobox genes in developing rostral brain. Nature 358, 687–690.
Simonnet, F., Celerier, M.L., Queinnec, E., 2006. Orthodenticle and empty spiracles genes are expressed in a segmental pattern in chelicerates. Dev. Genes Evol. 216, 467–480.
Snodgrass, R.E., 1935. Principles of Insect Morphology. McGraw Hill, New York.
St Johnston, D., Nusslein-Volhard, C., 1992. The origin of pattern and polarity in the Drosophila embryo. Cell 68, 201–219.
Telford, M.J., Thomas, R.H., 1998. Expression of homeobox genes shows chelicerate arthropods retain their deutocerebral segment. Proc. Natl. Acad. Sci. U. S. A. 95, 10671–10675.
Tessmar-Raible, K., Raible, F., Christodoulou, F., Guy, K., Rembold, M., Hausen, H., Arendt, D., 2007. Conserved sensory-neurosecretory cell types in annelid and fish forebrain: insights into hypothalamus evolution. Cell 129, 1389–1400.
The Tribolium Genome Sequencing Consortium, in press. The genome of the model beetle and pest Tribolium castaneum. Nature. doi:10.1038/nature06784.
Tomoyasu, Y., Denell, R.E., 2004. Larval RNAi in Tribolium Coleoptera for analyzing adult development. Dev. Genes Evol. 214, 575–578.
Treichel, D., Becker, M.B., Gruss, P., 2001. The novel transcription factor gene Sp5 exhibits a dynamic and highly restricted expression pattern during mouse embryogenesis. Mech. Dev. 101, 175–179.
Treichel, D., Schöck, F., Jäckle, H., Gruss, P., Mansouri, A., 2003. mBtd is required to maintain signaling during murine limb development. Genes Dev. 17, 2630–2635.

Urbach, R., Technau, G.M., 2003a. Molecular markers for identified neuroblasts in the developing brain of Drosophila. Development 130, 3621–3637.

Urbach, R., Technau, G.M., 2003b. Segment polarity and DV patterning gene expression reveals segmental organization of the Drosophila brain. Development 130, 3607–3620.

van der Zee, M., Berns, N., Roth, S., 2005. Distinct functions of the Tribolium zerknullt genes in serosa specification and dorsal closure. Curr. Biol. 15, 624–636.

van der Zee, M., Stockhammer, O., von Levetzow, C., Nunes da Fonseca, R., Roth, S., 2006. Sog/Chordin is required for ventral-to-dorsal Dpp/BMP transport and head formation in a short germ insect. Proc. Natl. Acad. Sci. U. S. A. 103, 16307–16312.

Vincent, A., Blankenship, J.T., Wieschaus, E., 1997. Integration of the head and trunk segmentation systems controls cephalic furrow formation in Drosophila. Development 124, 3747–3754.

Walldorf, U., Gehring, W.J., 1992. Empty spiracles, a gap gene containing a homeobox involved in Drosophila head development. EMBO J. 11, 2247–2259.

Weber, H., 1966. Grundriss der Insektenkunde. Gustav Fischer Verlag, Stuttgart.

Williams, N.A., Holland, P.W., 2000. An amphioxus Emx homeobox gene reveals duplication during vertebrate evolution. Mol. Biol. Evol. 17, 1520–1528.

Wimmer, E.A., Jäckle, H., Pfeifle, C., Cohen, S.M., 1993. A Drosophila homologue of human Sp1 is a head-specific segmentation gene. Nature 366, 690–694.

Wimmer, E.A., Simpson-Brose, M., Cohen, S.M., Desplan, C., Jäckle, H., 1995. Trans- and cis-acting requirements for blastodermal expression of the head gap gene buttonhead. Mech. Dev. 53, 235–245.

Wimmer, E.A., Frommer, G., Purnell, B.A., Jäckle, H., 1996. buttonhead and D-Sp1: a novel Drosophila gene pair. Mech. Dev. 59, 53–62.

Wimmer, E.A., Cohen, S.M., Jäckle, H., Desplan, C., 1997. buttonhead does not contribute to a combinatorial code proposed for Drosophila head development. Development 124, 1509–1517.

Wohlfrom, H., Schinko, J.B., Klingler, M., Bucher, G., 2006. Maintenance of segment and appendage primordia by the Tribolium gene knodel. Mech. Dev. 123, 430–439.

Wray, G.A., 2007. The evolutionary significance of cis-regulatory mutations. Nat. Rev., Genet. 8, 206–216.

3.2 Large-scale insertional mutagenesis of the coleopteran stored grain pest, the red flour beetle *Tribolium castaneum*, identifies embryonic lethal mutations and enhancer traps

In this part, the first large scale insertional mutagenesis screen in an insect outside *Drosophila* is described. The GEKU insertional mutagenesis screen was conducted as collaboration of four labs. This has identified a number of genes required for head development, segmentation and the development of appendages and some larval and pupal organs.

J. Trauner*, **J.B. Schinko***, M.D. Lorenzen*, T.D. Shippy*, E.A. Wimmer, R.W. Beeman, M. Klingler, G. Bucher, and S.J. Brown

* = co-first Authors

Authors contributions to the practical work:

J. Trauner*, M.D. Lorenzen*, T.D. Shippy*:	Analysis of 3045 lines for enhancer traps and determination of insertion site of 126 lethal and 9 sterile lines.
J.B. Schinko*:	Analysis of all Göttingen lines (2612) for enhancer traps in embryos, larvae, pupae and adults. Determination of the insertion site of almost all Göttingen lethal (233) and sterile (8) lines (for details see 3.2.2, 3.2.3 and 3.2.4)

Status: Published in BMC Biology 7: 73

3.2.1 Manuscript

Large-scale insertional mutagenesis of the coleopteran stored grain pest, the red flour beetle *Tribolium castaneum*, identifies embryonic lethal mutations and enhancer traps.

Jochen Trauner1*, Johannes B. Schinko2*, Marcé D. Lorenzen3*, Teresa D. Shippy4*, Ernst A. Wimmer#2, Richard W. Beeman3, Martin Klingler1, Gregor Bucher2, and Susan J. Brown4

*These authors contributed equally to this work

corresponding author

1Department of Biology, Developmental Biology, Friedrich-Alexander-University Erlangen, Germany

2Blumenbach Institute of Zoology and Anthropology, Dpt. of Dev. Biol., Georg-August-University Göttingen

3USDA-ARS-GMPRC, Manhattan, KS, USA

4Division of Biology, Ackert Hall, Kansas State University, Manhattan, KS, USA

Running head: Insertional mutagenesis in *Tribolium*

Keywords: *Tribolium castaneum*, *piggy*Bac, transposon, insertional mutagenesis, enhancer trap, jumpstarter

Corresponding author: Ernst A Wimmer

ABSTRACT

Given its sequenced genome and efficient systemic RNA interference response, *Tribolium castaneum* is a model organism well suited for reverse genetics. Even so, there is a pressing need for forward genetic analysis to escape the bias inherent in candidate gene approaches. To produce easy-to-maintain insertional mutations and to obtain fluorescent marker lines to aid phenotypic analysis, we undertook a large-scale transposon mutagenesis screen. In this screen, we produced more than 6,500 new *piggy*Bac insertions using an *in vivo* system. Of these, 421 proved to be recessive lethal, 75 were semi-lethal and 18 were recessive sterile, while 505 showed new enhancer-trap patterns. Insertion junctions were determined for 403 lines and often appeared to be located within transcription units. Insertion sites appeared to be randomly distributed throughout the genome, with the exception of a slight preference for reinsertion near the donor site. This collection of enhancer-trap lines and embryonic lethal lines has been made available to the research community and will foster investigations into diverse fields of insect biology, pest control, and evolution. Because the genetic elements used in this screen are species-nonspecific, and because the crossing scheme does not depend on balancer chromosomes, the methods presented herein should be broadly applicable for many insect species.

INTRODUCTION

During the past few years, the red flour beetle *Tribolium castaneum* has been developed into a powerful model organism suited for the study of short germ development, embryonic head and leg development, metamorphosis, cuticle metabolism, and other problems in insect biology. It is the first coleopteran species for which the genome sequence has become available (*Tribolium* genome sequencing consortium 2008). In-depth functional analysis of molecularly identified genes is enabled by the availability of germline transformation (Berghammer et al. 1999b, Lorenzen et al. 2003) and systemic RNA interference that is splice-variant-specific (Arakane et al. 2005) and feasible at all life stages (Brown et al. 1999; Bucher et al. 2002; Tomoyasu and Denell 2004). Furthermore, several tools and techniques have been developed that facilitate insertional mutagenesis in *Tribolium castaneum* (Horn et al. 2002, 2003; Pavlopoulos et al. 2004; Lorenzen et al. 2007). Although candidate gene approaches (reverse genetics) via RNA interference work very well in *Tribolium*, they are biased towards previously recognized genes and described mechanisms. In contrast, forward genetic approaches offer the opportunity to detect new gene functions not yet described in other model systems. Small-scale chemical mutagenesis screens have been performed in *Tribolium* (Sulston and Anderson, 1996; Maderspacher et al. 1998), but stock keeping of unmarked recessive mutants is difficult because of the ten chromosome pairs and the fact that less than half of the genome is covered by balancers (Berghammer et al. 1999a). In contrast, insertional mutagenesis screens using dominantly-marked "donor" transposons facilitate both stock keeping and gene identification.

Several species-nonspecific transposons including *Hermes, Minos,* and *piggy*Bac have been shown to function in *Tribolium* (Berghammer et al. 1999b; Pavlopoulos et al. 2004). Berghammer et al. (1999b) introduced EGFP under the control of the 3xP3 promoter as a universal, selectable marker for transgenic insects. This promoter is also responsive to nearby chromosomal enhancers (Lorenzen et al. 2003), allowing insertional mutagenesis to be combined with enhancer trapping (Horn et al. 2003). In our scheme, insertional mutagenesis is based on the controlled remobilization of a non-autonomous donor element stably inserted in the genome. The transposase needed to remobilize the donor element is provided by a helper element ("jumpstarter"). Lorenzen et al. (2007) created several jumpstarter strains using a modified *Minos* transposon to provide a source of *piggy*Bac transposase (Horn et al. 2003).

Here we report the first large-scale insertional mutagenesis screen conducted in an insect other than *Drosophila*. We have identified many insertion lines that are either homozygous lethal, homozygous sterile and/or show enhancer-trap patterns at various developmental stages. The genomic locations, enhancer-trap patterns (if present), recessive phenotypes, and genes affected by these transposon insertions are documented in the GEKU database (freely available at www.geku-base.uni-goettingen.de) and insertion lines are available upon request (GEKU: **G**öttingen, **E**rlangen, **K**SU, **U**SDA).

Our screening procedure should be applicable to many other insect species, because all genetic elements (transposons, promoters and marker genes) used in this screen are species-nonspecific (Horn et al. 2003). It also renders unnecessary the use of balancer chromosomes, which are not available for the vast majority of insect species. Obvious limitations may be the ability to rear the insect species in the laboratory, the feasibility of germline transformation to obtain donor and helper strains, and the ability to perform single-pair matings with high efficiency.

MATERIALS AND METHODS

Strains used: The "donor" strain used in this screen, Pig-19, carries a 3xP3-EGFP marked *piggy*Bac element that confers both insertion-site-independent, eye-specific EGFP expression and donor-site-dependent, muscle-specific EGFP expression (Lorenzen et al. 2003). We previously demonstrated that remobilization of the Pig-19 insertion results in G1 beetles lacking muscle-specific expression, but retaining eye-specific expression (Lorenzen et al. 2003; Lorenzen et al. 2007). Thus, the loss of muscle-specific expression can be used to detect remobilization events. The "jumpstarter/helper" strain used in this screen, M26, carries an X-chromosomal insertion of a 3xP3-DsRed marked Minos element (Lorenzen et al. 2007). Both strains are in a white-eyed pearl mutant background to facilitate detection of eye-specific fluorescence.

Generating new *piggy*Bac insertions: We used a P1, P2 and F1 to F4 scheme to comply with standard *Drosophila* F1 and F3 genetic screens, respectively. Donor remobilization first occurred in the germline of the P2 generation, while new insertions and mutant homozygotes first appeared in the F1 and F3 generations, respectively. All crosses were carried out at 30–32°C. Virgin females were collected as pupae and stored at 23°C for up to six weeks prior to use. Insertional mutagenesis is described in detail in Lorenzen et al. (2007). In summary, P1 mass-matings were set up between donor males and helper females (Fig. 1A) and subcultured at intervals of 2–3 weeks. P2 offspring were collected as pupae and examined to verify the presence of both donor (EFGP marker) and helper (DsRed marker) constructs. Individual P2 virgin females were outcrossed to three pearl males each to ensure insemination (Fig. 1B). New insertions were recognized in the F1 progeny by the loss of donor-site-dependent EGFP expression (i.e. muscle fluorescence) coupled with retention of insertion-site-independent EGFP expression (i.e. eye fluorescence). For each P2 outcross, a single F1 beetle carrying a new insertion was outcrossed once again to pearl to verify a single insertion (based on 50% Mendelian segregation of the new insert) and to generate families for subsequent analysis (Fig. 1C). Additionally, depending on the new chromosomal location of piggyBac, a new insertion might show a novel enhancer-trap pattern. Even when several beetles with new insertions among the offspring of one P2 cross were found, only one beetle was chosen for continued study in order to ensure independent origin of each new insertion. This was necessary because several offspring carrying the same insertion could appear within a P2 family as a result of a premeiotic remobilization event. For each F1 outcross, five female and three male F2 siblings were crossed to each other to establish new insertion strains and to enable testing for homozygous viability (F2 cross; Fig. 1D). To accomplish the latter, we performed a number of single-pair F3 matings (Fig. 1E) and analysed their progeny for the presence of the donor element (see below).

Statistical considerations: If an insertion mutant is homozygous viable, then (after positive marker selection) the progeny of an F2 cross will consist of a 1:2 ratio of homozygous to heterozygous beetles. Under the assumption of random sib-mating, 11.1% (1/3 x 1/3) of all F3 single-pair matings would have been crosses between two homozygous beetles, 44.4% [2x (1/3 x 2/3)] between one homozygous and one heterozygous beetle, and 44.4% (2/3 x 2/3) between two heterozygous beetles. This implies that about 55.5% (11.1% + 44.4%) of all single-pair matings with a fully viable insertion were expected to produce only EGFP-expressing progeny (because at least one parent was homozygous for the insertion). The remaining 44.4% were expected to produce mixed progeny (i.e. ~75% EGFP positive and ~25% EGFP negative) because both parents were heterozygous for the insertion. In contrast, for recessive lethal insertions, no homozygous beetles would have been present in the F3 generation and all F3 crosses would produce mixed progeny. Thus, the presence of at least one F4 animal lacking the transformation marker (i.e. with nonfluorescent eyes) indicated heterozygosity of both parents. Absence of EGFP-negative progeny indicated homozygosity of at least one of the parents. Depending on the distribution of phenotypes in the offspring, each single-pair mating was assigned to one of five categories (see Table 1 for details).

Since more than 40% of all single-pair matings were expected to produce mixed progeny, even if the insertion is fully viable, we analyzed a total of 20 single-pair matings before concluding that an insertion was lethal. On the other hand, since viable insertions usually could be identified after evaluating just a few single-pair matings we split the 20 crosses into two consecutive rounds to maximize throughput. The second round of single-pair matings was set up only if an insertion was not clearly identified as being viable after evaluating the single-pair matings from the first round (Table 2).

Potential errors that could occur during this test for recessive lethality are: (1) A homozygous-viable insertion mutant could be falsely judged homozygous lethal because all single-pair matings produced mixed progeny. This could occur if, by chance, all single-pair matings consisted of heterozygous beetles. The probability of such an occurrence is $(2/3)^n$ (n=number of beetles tested), because two-thirds of all marked F3 beetles carrying a viable insertion are heterozygous. For eight single-pair matings (number of test beetles = 16), this probability equals 0.15%. For 20 single-pair matings, the probability that all (40) test beetles selected from a homozygous-viable line are heterozygous is only 9.0×10^{-6}. Thus, evaluating 20 single-pair matings is sufficient to exclude false-positive lethal lines with a very high level of confidence. (2) A homozygous-lethal insertion (all F3 beetles are heterozygous) could be falsely identified as homozygous viable if, by chance, no progeny lacking the EGFP marker are observed from a single-pair mating, even though 25% are expected. The probability of this happening when 20 progeny are analyzed is about 0.3% (0.75^n; n=number of progeny screened). Because the probability of misdiagnosing a lethal insertion rises if fewer progeny are analyzed, single-pair matings yielding a total of fewer than 20 progeny were not used to make inferences about the lethality of the insertion ("uninformative single-pair mating" in Table 1) unless some of the progeny were non-EGFP.

Overcoming a negative X-chromosome bias: The fact that the helper insertion used in this work is X-linked imposed restrictions on the design of our crossing scheme. X-chromosomal insertions that were homozygous lethal or sterile could be obtained only if the following is considered: Because only new transformants that segregated away from the helper element were selected, hybrid females had to be used to set up P2 crosses in order to avoid bias against new X-linked insertions. Additionally, males with a new hemizygous X-linked lethal insertion would not survive and ones hemizygous for a new X-linked sterile insertion would be useless for generating a new stock. Hence, one could obtain X-linked lethal and sterile insertions only if female beetles carrying the donor element were used to set up the P2 as well as the F1 cross. Therefore, we selected only female hybrids and used females carrying new insertions whenever possible.

Efficiency of detecting new insertions: At least one new insertion was detected in about 30% of all P2 crosses when about 20 offspring were screened. The percentage of P2 crosses that yield new insertions can be greatly increased by screening a larger number of progeny per P2 cross. For a subset of P2 crosses we screened 100 progeny per cross, and found at least one new transformant in every case. In practice, about 10–30 P2 offspring at the pupal stage were present when the P2 progeny were screened for new transformants. The decision to discard the larval offspring of a P2 family when no new insertion could be detected in the first attempt represented a compromise between the aim to obtain a large number of independent insertions and the need to maximize the likelihood of finding at least one insertion in each family.

Determination of insertion sites: The genomic location of an insert was determined by sequencing flanking DNA obtained by one of the following three PCR-based methods: inverse PCR (Ochman et al. 1988), universal PCR (Beeman and Stauth 1997; Lorenzen et al. 2003), or vectorette PCR (Arnold and Hodgson 1991). The procedure for inverse PCR including primer design was adapted from "Inverse PCR and Sequencing Protocol on 5 Fly Preps", Exelixis Pharmaceutical Corp (Thibault et al. 2004). Following DNA isolation, approximately 1 µg of DNA was digested with Sau3A1, BfUC1, or Ase1 (for 5' iPCR) or HinP1 (for 3' iPCR). Approximately 100 ng of digested DNA was then self-ligated to obtain circular DNA fragments, followed by two rounds of nested PCR. DNA templates (PCR products and/or cloned PCR products) were sequenced by Seqlab (Göttingen), Macrogen (Corea), or using an ABI 3730 DNA sequencer (Sequencing and Genotyping Facility, Plant Pathology, Kansas State University). Data analysis was performed using Vector NTI® software (Invitrogen). After trimming vector sequences, flanking DNA sequences were then searched (Blastn) against *Tribolium castaneum* genome sequences at HGSC, Baylor College of Medicine (http://www.hgsc.bcm.tmc.edu/projects/*tribolium*/), NCBI (http://www.ncbi.nlm.nih.gov/genome/seq/BlastGen/BlastGen.cgi?taxid=7070) or BeetleBase (http://beetlebase.org/). If the insertion was in a GLEAN gene prediction, a transcription unit or region indicated by *Drosophila* Blast or other gene prediction method as a potential gene, the predicted *Tribolium* gene was examined by blast analysis at FlyBase for the top *Drosophila* hit, and NCBI (nr database) to identify other potential orthologs.

Medea (maternal effect dominant embryonic arrest): The offspring of crossings between hybrid females and pearl males (P2 generation) showed a severe segregation distortion: The number of EGFP-expressing beetles was ~98% (instead of 50%), whereas the DsRed marker showed the expected 1:1 ratio and segregated independently of the EGFP marker. The unexpected segregation ratio of EGFP is due to close linkage (~2cM) of the maternally acting selfish gene Medea (Beeman et al. 1992) and the Pig-19 insertion (Lorenzen et al. 2007) on LG3. However, the segregation ratios of new insertions were affected only when the *piggy*Bac element reinserted near the original donor insertion (representing a local hop).

GEKU-base: All available information about the analyzed insertion lines can be found at a web-based database called GEKU-base (www.geku-base.uni-goettingen.de). Information provided includes (if available): picture and descriptions of enhancer traps and phenotypes, flanking sequences and chromosomal location, affected genes and their orthologs. GEKU-base also provides information on how to obtain insertion lines.

EGFP and DsRed analysis:

Marker-gene fluorescence was detected using either a Nikon fluorescence stereomicroscope SMZ1500 [at G and E], an Olympus SZX12 fluorescence stereomicroscope (Olympus Corporation, Tokyo, Japan), or a Leica MZ FLIII fluorescence stereomicroscope (Leica Microsystems Inc., Wetzlar, Germany). The filter sets used for EGFP expression were: [G= 470/40 nm excitation filter, a 500 nm LP emission filter, and a 495 nm beamsplitter] [E= 480/40 nm excitation filter, a 510 nm emission filter, and a 505 nm beamsplitter] [K= 480/40 nm excitation filter and a 535/50 nm emission filter] [U= GFP Plus filter set (excitation filter: 480/40 nm, barrier filter: 510 nm)]. The filter sets used for DsRed expression were: [G= 546/12 nm excitation filter, a 605/75 nm emission filter, and a 560 nm beamsplitter] [E= 565/30 excitation filter, a 620/60 nm emission filter, and a 585 nm beamsplitter] [K= 545/30 excitation filter and a

620/60 emission filter] [U = TXR TEXAS RED filter set (excitation filter: 560/40 nm, barrier filter: 610 nm)]. To detect enhancer-trap patterns in embryos, we dechorionated embryos derived from F3-crosses.

RESULTS

Test for lethality and sterility: Following the procedure illustrated in Figure 1, a total of 6816 new, independently derived insertions were isolated in the F1 generation and of these, 5657 new insertion lines were successfully tested for lethality/sterility. 589 potentially homozyogous lethal lines were identified in a first round of the F3 crosses, of which 421 were confirmed to be homozygous lethal in a second round (Table 3, for details on the two rounds of screening F3 crosses please see materials and methods). A subset of the viable insertions lines that produced a smaller number of homozygotes than expected were tested for semi-lethality. Insertion lines were designated as potentially semi-lethal if only one single-pair mating in the first round of F3 crosses or fewer than four single-pair matings after the second round indicated homozygosity of a parent. This was true for 236 insertions (out of the subset of 2940 insertions analyzed in G and E) after the first round, of which 75 remained in this category after the second round. Hence, 2.5% (75/2940) of all insertions tested for semi-lethality met the criteria for semi-lethality. This somewhat relaxed scoring criterion reduced the likelihood of missing or overlooking lethal or semi-lethal mutations.

Potentially homozygous sterile insertions lines were identified by evaluating the single-pair matings: Whenever two or more of the initial single-pair F3 self-crosses (round one) (Fig. 1E) failed to produce offspring (although the parents were alive and healthy), the line was classified as potentially sterile. This was true for 160 insertions (Table 3). We used either of two methods to confirm or refuse a tentative diagnosis of recessive sterility. In the first method, we set up a second round of single-pair self-crosses bringing the total number of F3 crosses to 20. The diagnosis was considered to be corroborated when the number of single-pair matings not producing any offspring increased to four or more. Using this definition, 124 potentially sterile lines were reduced to 21. However, further testing of these sterile insertion lines showed that this criterion was not always reliable. In the second method we set up 10 male and 10 female outcrosses. The diagnosis of recessive sterility was considered to be corroborated if either no fertile homozygous female or no fertile homozygous male could be identified in these outcrosses. Out of 36 potentially sterile lines tested by the second method, only eight lines fulfilled this definition of sterility. Since the second follow-up test appeared to be more rigorous than the first, we retested 11 of the 21 apparently sterile lines from the former test using the more rigorous criterion. All 11 lines proved to be fertile in both sexes. It seems to be clear that most sterile lines found by using the first criterion are false-positives. Hence, we suggest using the stricter test for recessive sterility, which has the added benefit of identifying the affected sex.

Detection of enhancer traps: We analyzed all new insertion lines for enhancer-dependent EGFP expression, and detected novel patterns at all developmental stages. Although we observed a bias for certain patterns (i.e. certain CNS patterns, segmentally-repeated stripes in embryos or small dots at the hinges of extremities in larvae and pupae), we identified 505 unique enhancer-trap patterns. 3xP3-driven EGFP expression is typically seen in only the eyes and central nervous system (Lorenzen et al. 2003). However, the bias for certain patterns might be caused by a favoured expression in certain tissues due to the paired-type homeobox binding sites in the 3xP3 element of the transformation marker (Horn et al. 2000). For a random subset of about 200 of all newly identified insertions, we also dissected pupae and adults to look for EGFP expression in internal organs that might not be visible without dissection. Such expression patterns (e.g. a spermatheca enhancer) were found only rarely. Examples of enhancer-trap lines are shown in Figure 2A-H. Descriptions and/or pictures of all enhancer-trap lines together with information about their chromosomal locations (when known) are available in GEKU-base (www.geku-base.uni-goettingen.de; see Materials and Methods).

Analysis of lethal lines and developmental phenotypes: We analyzed the embryonic cuticle phenotypes of many lines identified as lethal and found a number of distinct cuticular abnormalities (Figure 2I-L). For example, line G08519 displays a phenotype similar to the proboscipedia ortholog maxillopedia in that maxillary (grey arrows) and labial (white arrow) palps are transformed to legs (Fig. 2J; Shippy et al. 2000a; 2000b). Indeed, the insertion is located in the first intron of maxillopedia. In addition, many lethal lines showed a high proportion of embryos that died prior to cuticularization, indicating early embryonic lethality.

To test whether the "semi-lethal" lines are false positives or true lethals with occasional "escapers", we checked what portion of these lines (Göttingen subset) produce lethal L1 cuticle phenotypes (at least two cuticles with similar strong defects in one preparation when scoring at least 10 individuals) and compared it to the percentage of cuticle phenotypes produced by the other classes. 25.8% (8/31) of a random selection of lines complying with the strict definition of lethality showed such phenotypes. Of lines with one or two single-pair matings (out of 20) indicating homozygosity (semilethality), this portion was 16.6% in each case (5/30 and 3/18, respectively). Lines with three single-pair matings indicating homozygous viability gave rise to cuticle phenotypes in only 6.25% (1/16). Thus analyzing "semi-lethal" lines led to the identification of additional cuticle phenotype-inducing mutations.

Chromosomal location of new *piggy*Bac insertions: We determined the chromosomal location for 400 *piggy*Bac insertions by BLAST analysis of the isolated flanking sequence against the *Tribolium* genome (see materials and methods). These insertions included lethal, semi-lethal and sterile as well as viable lines that showed an enhancer-trap pattern. The distribution of 280 homozygous lethal insertions on the linkage groups is shown in Figure 3. The lethal insertions appear to be distributed randomly among the linkage groups, showing a range from 1.1 insertions per Mb for linkage group 10 up to 2.2 insertions per Mb for linkage group 4 (Table 4). Superimposed on the generally random pattern of insertion site locations, there appear to be insertion hotspots and coldspots, the most evident example being the hotspot for local reinsertion near the donor site on linkage group 3. The largest region devoid of any insertions was a region spanning 3.4 Mb on linkage group 3 (Figure 3).

DISCUSSION

The GEKU insertional mutagenesis screen was designed to meet the following criteria: It should be rapid and simple (i.e. involve as few generations as possible); and the analysis of the resulting insertion lines should be highly efficient (i.e. producing only a small number of false positive lethal or sterile lines while also minimizing the frequency of false negatives; see Material and Methods).

Large-scale insertional mutagenesis is feasible in a coleopteran species: Based on a pilot screen published in Lorenzen et al (2007) we have performed the first high-efficiency, large-scale insertional mutagenesis screen in an insect species outside the genus *Drosophila*, and we have established a crossing scheme that circumvents the need for balancer chromosomes or embryo injections. From our experience, we estimate that using the procedure presented here, one person should be able to establish 150 lethal strains per year. While the GEKU screen has identified many interesting enhancer traps and lethal phenotypes, genome-wide saturation would be difficult to achieve at the current level of efficiency. The most time-consuming step is setting up and evaluating 20 single-pair matings for each new insertion line to detect recessive lethality. For this reason we set up a small number of single-pair matings first, as most viable insertions can be identified by evaluating just a few crosses from each subset. However, it was important to assess the fertility of all remaining single-pair matings in order to ensure that recessive sterile insertions were detected.

Lethal insertions are readily detected while insertions causing sterility are difficult to detect: We found that lethal lines were readily detected by single pair matings. Based on the frequency with which semi-lethal lines produced strong L1 larval cuticle phenotypes, we suggest defining lines as potentially lethal when only one or two out of 20 single-pair matings indicate homozygosity. However, our definition of sterility proved to be too lax in the beginning, since most potentially sterile lines turned out to be false-positives in more detailed analysis that also allowed us to determine the gender in which the sterility occured.

Comparing efficiencies with *Drosophila melanogaster* insertional mutagenesis and enhancer trap screens: The efficiency of generating lethal mutations by *piggy*Bac-based insertional mutagenesis in *Tribolium* (7.4%) is similar to equivalent screens in *Drosophila* based either on *piggy*Bac (Horn et al. 2003; Häcker et al. 2003) or P elements (Cooley et al. 1988; Bellen 1999; Peter et al. 2002). Whether the efficacy of such screens can potentially be doubled by the inclusion of splice acceptor sites or insulator sequences within the mutator element – as has been shown in *Drosophila* (Thibault et al. 2004) – still has to be determined in *Tribolium*.

The enhancer detection rate within this large scale insertional mutagenesis screen was also 7.4%. This is actually higher than in a comparable *Drosophila* screen where enhancer detection without a suitable amplification system was about 2% (Horn et al. 2003). Only after including an ectopic expression system-based amplifier system could *Drosophila* enhancer detection be raised to 50% (Horn et al. 2003). However, such directed expression systems still need to be further developed and assayed in *Tribolium* before they can be used in insertional mutagenesis screens.

Correlation of phenotype (lethality, sterility, enhancer trap) with insertion site proximity to protein coding sequences (CDS): In 14% of all lethal insertions, *piggy*Bac had clearly jumped into the coding sequence of a gene. However, the majority of lethal insertions (61%; see Table 5) were located in introns, apparently disrupting transcription or splicing of the affected gene. One possibility is that the SV40 UTR in the transposon, which serves as a terminator of transcription in both directions, causes early transcriptional termination of the host gene.

Ways to enhance overall efficiency: In the described scheme, when new crosses had to be set up, one had to switch between fluorescence (to identify the transformation marker) and normal light (to identify the sex of the pupae) several times, which was a time-consuming procedure. To improve this situation considerably, we constructed and are testing new donors that use rescue of eye color by *vermilion* as an indication of transformation (Lorenzen et al. 2002a; Lorenzen et al. 2002b). The use of such a system will also facilitate stock-keeping. Another way to enhance screening efficiency might be the establishment of donors that include an artificial maternal-effect selfish element (e.g. MEDEA). This would be an elegant means to enhance both generation of lethal lines and their stock keeping (Chen et al. 2007).

Acknowledgments

We are deeply indebted to Elke Küster, Susanne Fischer, Tatum Kimzey, Kathy Leonard and Jonna Voorhees for technical assistance in production and characterization of transgenic lines. All programs and services of the US Department of Agriculture are offered on a nondiscriminatory basis, without regard to race, color, national origin, religion, sex, age, marital status or handicap. Mention of trade names or commercial products in this publication is solely for the purpose of providing specific information and does not imply recommendation or endorsement by the U.S. Department of Agriculture. This work has been funded by the United States Department of Agriculture (USDA) within the CSREES_NRI program "Functional Genomics of Agriculturally Important Organisms – Insects and Mites" grant #2004-35604-14250 (EAW, RWB, MK, SJB).

FIGURE LEGENDS

Figure 1

Screening procedure for the creation of new insertions and selection of lethal and sterile insertions. (A) P1 cross: Mass-crosses were made between the donor strain carrying the *piggy*Bac element pBac[3xP3-EGFPaf] and the helper strain carrying the *Minos* element Mi[3xP3-DsRed; Dm'hsp70-pBac]. Both strains are marked by eye-specific EGFP and DsRed expression, respectively. Note the additional enhancer-trap pattern of the donor strain. (B) P2 cross: Single hybrid females carrying both, the donor and the helper element (simultaneous expression of EGFP and DsRed in the eyes), are crossed to three pearl males. *piggy*Bac can be remobilized by the activity of the transposase. (C) F1 cross: A single individual carrying a new insertion is selected out of the offspring of a P2 cross and crossed to several pearl mates. A remobilization event is evident in beetles that still show marker gene expression in the eyes, but have lost the muscle-specific enhancer. Please note the altered enhancer-trap phenotype of the new insertion line in this example (EGFP expression in the wings). Only individuals carrying a new insertion that did not inherit the helper element (i.e. no DsRed expression in the eyes visible) were chosen. (D) F2 cross: All offspring of the F1 cross showing marker gene expression are heterozygous for the insertion and are sibling-crossed to establish a strain. (E) F3 cross: Several single-pair matings are set up. (F) Test for lethality and sterility: Marker gene expression of the offspring of each single-pair mating is evaluated to determine whether their parents were hetero- or homozygous for the *piggy*Bac insertion (see Materials and Methods). Each single-pair mating is assigned to one out of five classes (small black arrows; see also Table 1). The combined evaluation of all single-pair matings was used to define the phenotype of the insertion (see Table 2 for further details). (G) Summary on all identified lethal, semi-lethal and sterile insertions, as well as insertions that show an enhancer trap (see text and Table 3 for details).

Figure 2

Enhancer traps (A-H) and mutant phenotypes (I-L) of *piggy*Bac insertion lines. Gene names refer to respective *Drosophila* orthologs. (A) The line E00321 carries an insertion in *lethal (2) giant larvae* and shows EGFP expression in the cuticle during all larval stages. This line is homozygous lethal. (B) The line E00713 carries an insertion 149 bp upstream of the 5' end of Glean_03347, (*Glutathione S transferase*) and displays EGFP expression in a subset of somatic muscles. This line is homozygous viable. (C) The line G01004 carries an insertion near *Ultrabithorax* and shows EGFP expression in the abdomen. This line is homozygous viable. (D) The line G04717 carries an insertion near *lame duck*; EGFP is expressed in two lateral stripes, which based on the similarity to the *Drosophila* expression pattern is presumably locatetd in the mesoderm. This line is homozygous viable. (E) The line KT1539 inserted near the gene *pointed*; EGFP is localized in a salt and pepper pattern in the ventral abdominal epidermis; this line is homozygous lethal. (F) KS030 has an insertion in an intron of *lozenge*. EGFP expression is detected in the embryonic distal legs. This line is homozygous lethal. (G) KS406 carries an insertion in an intron of GLEAN_00277 which shows identity to protein tyrosine phosphatases Other genes in the vicinity of this insertion are *Fgf8* and *Or48*; EGFP is expressed in the embryonic hindgut and in segmental stripes; the line is homozygous viable. (H) MH30a has an insertion near *female sterile (2) Ketel*. EGFP expression is in the proximal embryonic leg, as well as in T2, T3, and A1 spots, and a posterior ring in the first instar larva. This line is homozygous viable. (I) The line E00916 carries an insertion in an exon of GLEAN_08270 (*Drosophila* ortholog: *Cyclin D*). Homozygous embryos are poorly differentiated and have bubbly short legs and segmental defects. This line is homozygous lethal. (J) The insertion G08519 is located in the first intron of *proboscipedia*; the phenotype corresponds to the one described for *Tc-maxillopedia* mutants: maxillary (grey arrows) and labial (white arrow) palps are transformed to legs while the overall morphology of the segments appears unchanged. (K) The phenotype of KT1096 is characterized by dorsal cuticular defects and possibly additional patterning or differentiation problems. This insertion is in an intron of the *pecanex* ortholog. (L) Embryos homozygous for E03501 develop rudimentary appendages in the 1st abdominal segment. This insertion is in an intron of the *Tribolium* ortholog of *Ftz-F1*.

Figure 3

Distribution of 280 lethal *piggy*Bac insertions on linkage groups 1 to 10. Location of the donor line Pig-19 on LG3 is indicated. Detail of LG3 magnified 12x. Scale bar = 1 Mb.

Table 1: Evaluation of F3 single-pair matings

Offspring of a single-pair mating	Interpretation/Result
No offspring, but parents were still alive at the time of evaluation	This indicates sterility of at least one of the parents
No offspring, but one or both parents were dead at the time of evaluation	*uninformative single-pair mating**
At least one of the progeny was EGFP negative	This indicates heterozygosity of both parents.
All progeny were EGFP positive, at least 20 beetles present	This indicates homozygosity of one or both parents
All progeny were EGFP positive, but less than 20 beetles present	*uninformative single-pair mating**

*These single-pair matings were omitted from the overall evaluation (see Materials and Methods)

Table 2: Test for lethality and sterility

First round of F3 single-pair matings (SPM)   **Second round of F3 single-pair matings (SPM)**

Offspring	Phenotype	Offspring	Phenotype
At least one SPM indicates homozygosity	Viable	At least one SPM (in total) indicates homozygosity	Viable
All informative[§] SPM indicate heterozygosity of both parents	Potentially lethal	All informative[§] SPM indicate heterozygosity of both parents	Lethal
At least two SPM without any offspring but with living parents	Potentially sterile	Unable to find at least four SPM (in total) without any offspring but with living parents (method 1) OR unable to identify either a fertile homozygous female or a fertile homozygous male (method 2)	Sterile

After the first round of single-pair matings, all viable insertions were discarded (unless an enhancer traps was detected). All potentially lethal and potentially sterile lines were restested in a second round of single-pair matings.

[§]A single-pair mating (SPM) is uninformative if it produces no offspring and one or both parents are dead, or less than 20 offspring are present and all of them are GFP-positive (see Table1 and Materials and Methods)

Table 3: Results of lethality/sterility test (F3 cross)

First round of single-pair matings		Second round of single-pair matings	
phenotype	number of insertions	phenotype	number of insertions
viable[a]	4908 (86,8%)	viable	250 (4.4%)
potentially lethal	589 (10.4%)	lethal	421 (7.4%)
potentially sterile	160 (2.8%)	sterile[b]	18 (0.3%)
		not retested	60 (1.1%)

A total of 5657 lines (100% for all numbers given) were tested for potential lethality or sterility by a first round of single pair matings (left half of table; see results and M&M for details). Those that matched the criteria (749/5657) were retested by a second round of additional single pair matings in order to eliminate false positives (right half of table). Only those lines that matched the definition in the second round were considered to be lethal or sterile.

[a] This number includes 236 lines that were considered potentially semi-lethal (see text for definition of semi-lethality). Because this was done only on a subset of 2940 lines, the numbers are not given separately.

[b] For eight of these 18 lines sterility was confirmed by out-crossing, 10 were detected by single-pair matings and were not retested.

Table 4: Chromosomal location of lethal piggyBac insertions

Chromosome	Insertions	Chromosome size* [Mb]	Insertions / Mb
X	19	10.9	1.7
2	31	20.2	1.5
3	67	39.0	1.7
4	30	13.9	2.2
5	31	19.1	1.6
6	18	13.2	1.4
7	33	20.5	1.6
8	22	18.0	1.2
9	27	21.5	1.3
10	13	11.4	1.1
unmapped	14		

305 insertions were localized in the genome sequence. Of these, 14 were on unmapped scaffolds and 11 could be assigned to chromosomes, but not to the exact position. The distribution of the remaining 280 lethal insertions in the genome is shown in Fig. 3.

*based on NCBI map viewer, build 2.1.

Table 5: Detailed analysis of lethal *piggy*Bac insertion sites

location	number	%
Intron	185	61
CDS*	42	14
< 500 bp**	27	9
500 bp - 2500 bp**	24	8
distant (> 2500 bp)**	27	9
all localized	305	100
seq or blast problem	54	
not sequenced	62	
all lethal	421	

*exons excluding UTRs

**distance to next gene

REFERENCES

Arakane, Y., S. Muthukrishnan, K. J. Kramer, C. A. Specht, Y. Tomoyasu, M. D. Lorenzen, M. Kanost, and R. W. Beeman, 2005 The Tribolium chitin synthase genes TcCHS1 and TcCHS2 are specialized for synthesis of epidermal cuticle and midgut peritrophic matrix. Insect Mol Biol 14: 453–463.

Arnold, C., and I. Hodgson, 1991 Vectorette PCR: a novel approach to genomic walking. PCR Meth Appl 1: 39–42.

Beeman, R. W., K. S. Friesen, and R. E. Denell, 1992 Maternal-effect selfish genes in flour beetles. Science 256: 89–92.

Beeman, R. W., and D.M. Stauth, 1997 Rapid cloning of insect transposon insertion junctions using 'universal' PCR. Insect Mol Biol. 1: 83–88.

Bellen, H. J. 1999 Ten years of enhancer detection: lessons from the fly. Plant Cell. 11: 2271–2281.

Berghammer, A. J., G. Bucher, F. Maderspacher, and M. Klingler, 1999a A system to efficiently maintain embryonic lethal mutations in the flour beetle Tribolium castaneum. Dev Genes Evol. 209: 382–389.

Berghammer, A. J., M. Klingler, and E. A. Wimmer, 1999b A universal marker for transgenic insects. Nature 402: 370–371.

Brown, S. J., J. P. Mahaffey, M. D. Lorenzen, R. E. Denell, and J. W. Mahaffey, 1999 Using RNAi to investigate orthologous homeotic gene function during development of distantly related insects. Evol Dev 1: 11–15.

Bucher, G., J. Scholten, and M. Klingler, 2002 Parental RNAi in Tribolium (Coleoptera). Curr. Biol. 12: R85–86.

Chen, C.-H., H. Huang, C. M. Ward, J. T. Su, L. V. Schaeffer et al., 2007 A synthetic maternal-effect selfish genetic element drives population replacement in Drosophila. Science 316: 597–599.

Cooley, L., R. Kelley, and A. Spradling, 1988 Insertional mutagenesis of the Drosophila genome with single P elements. Science. 239: 1121–1128.

Häcker, U., S. Nystedt, M. P. Barmchi, C. Horn, and E. A. Wimmer, 2003 piggyBac-based insertional mutagenesis in the presence of stably integrated P elements in Drosophila. Proc Natl Acad Sci U S A. 100: 7720–7725.

Horn, C., and E. A. Wimmer, 2000 A versatile vector set for animal transgenesis. Dev Genes Evol 210: 630–637.

Horn, C., B. G. M. Schmid, F. S. Pogoda, and E. A. Wimmer, 2002 Fluorescent transformation markers for insect transgenesis. Insect Biochem. Mol. Biol. 32: 1221–1235.

Horn, C., N. Offen, S. Nystedt, U. Häcker, and E. Wimmer, 2003 piggyBac-Based insertional mutagenesis and enhancer detection as a tool for functional insect genomics. Genetics 163: 647–661.

Lorenzen, M. D., S. J. Brown, R. E. Denell, and R. W. Beeman, 2002a Cloning and characterization of the Tribolium castaneum eye-color genes encoding tryptophan oxygenase and kynurenine 3-monooxygenase. Genetics 160: 225–234.

Lorenzen, M. D., S. J. Brown, R. E. Denell, and R. W. Beeman 2002b Transgene expression from the Tribolium castaneum Polyubiquitin promoter. Insect Mol Biol. 11: 399–407.

Lorenzen, M. D., A. J. Berghammer, S. J. Brown, R. E. Denell, M. Klingler, and R. W. Beeman, 2003 piggyBac-mediated germline transformation in the beetle Tribolium castaneum. Insect Mol. Biol. 12: 433–440.

Lorenzen, M. D., T. Kimzey, T. D. Shippy, S. J. Brown, R. E. Denell, and R. W. Beeman, 2007 piggyBac-based insertional mutagenesis in Tribolium castaneum using donor/helper hybrids. Insect Mol. Biol. 16: 265–275.

Maderspacher, F., G. Bucher, and M. Klingler, 1998 Pair-rule and gap gene mutants in the flour beetle Tribolium castaneum. Dev Genes Evol. 208: 558–568.

Ochman, H., A. S. Gerber, and D. L. Hartl, 1988 Genetic applications of an inverse polymerase chain reaction. Genetics 120: 621–623.

Pavlopoulos, A., A. J. Berghammer, M. Averof, and M. Klingler, 2004 Efficient Transformation of the Beetle Tribolium castaneum Using the Minos Transposable Element: Quantitative and Qualitative Analysis of Genomic Integration Events. Genetics 167: 737–746.

Peter, A., P. Schöttler, M. Werner, N. Beinert, G. Dowe, et al., 2002 Mapping and identification of essential gene functions on the X chromosome of Drosophila. EMBO Rep. 3: 34–38.

Shippy, T. D., S. J. Brown, and R. E. Denell, 2000a Maxillopedia is the *Tribolium* ortholog of proboscipedia. Evol Dev. 2: 145–151.

Shippy, T. D., J. Guo, S. J. Brown, R. W. Beeman, and R. E. Denell, 2000b Analysis of maxillopedia expression pattern and larval cuticular phenotype in wild-type and mutant *Tribolium*. Genetics 155: 721–731.

Sulston, I. A., and K. V. Anderson, 1996 Embryonic patterning mutants of *Tribolium castaneum*. Development 122: 805–814.

Thibault, S. T., M. A. Singer, W. Y. Miyazaki, B. Milash, N. A. Dompe, et al., 2004 A complementary transposon tool kit for Drosophila melanogaster unsing P and piggyBac. Nature Genetics 36: 283–287.

Tomoyasu, Y., and R. E. Denell, 2004 Larval RNAi in *Tribolium* (Coleoptera) for analyzing adult development. Dev Genes Evol 214:575–578.

Tribolium Genome Sequencing Consortium, 2008 The genome of the model beetle and pest *Tribolium castaneum*. Nature 452(7190): 949–955.

Insertional mutagenesis screen-Manuscript

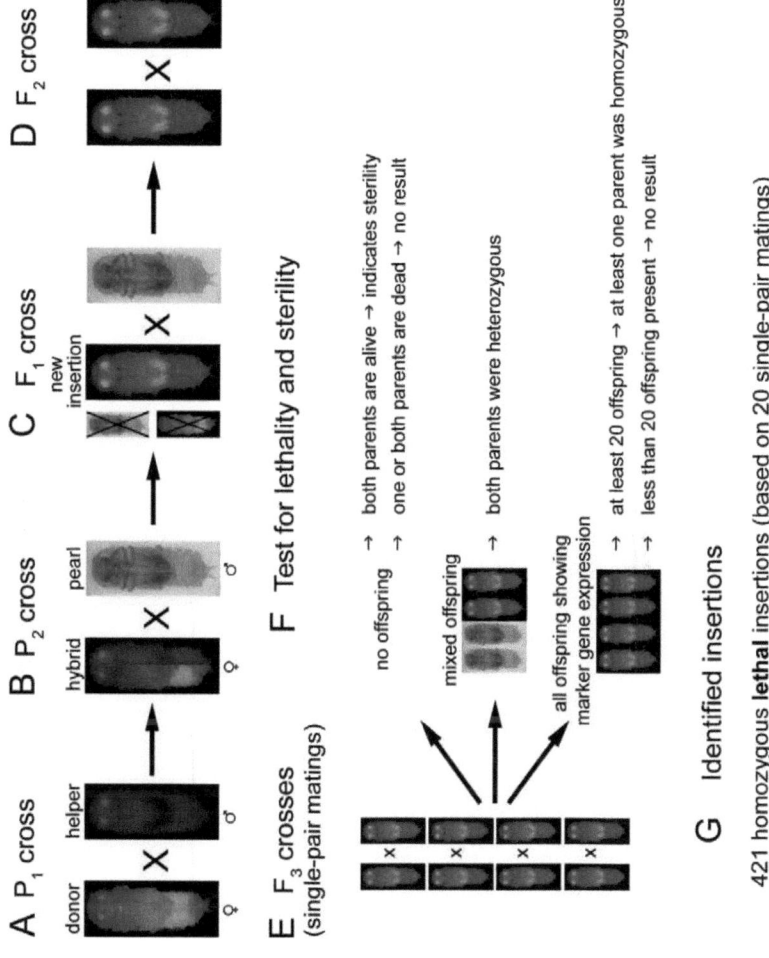

Figure 1

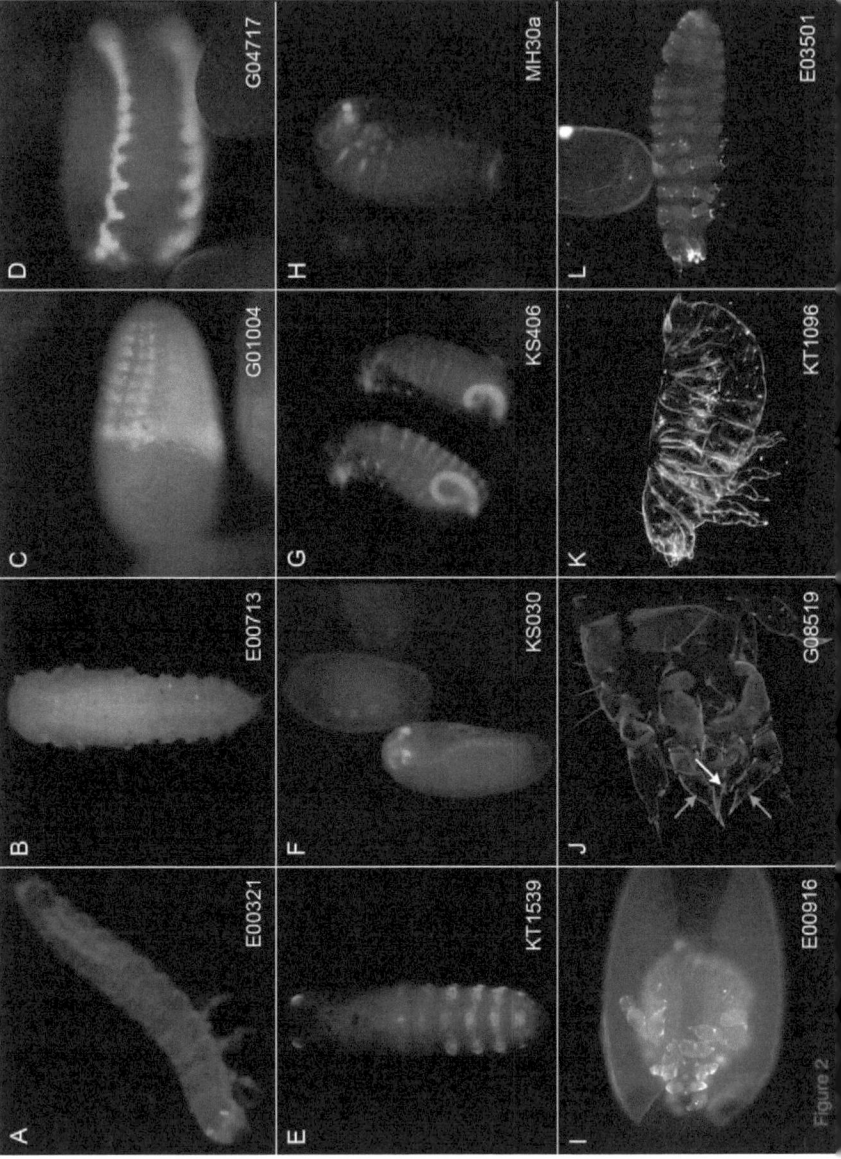

Insertional mutagenesis screen-Manuscript

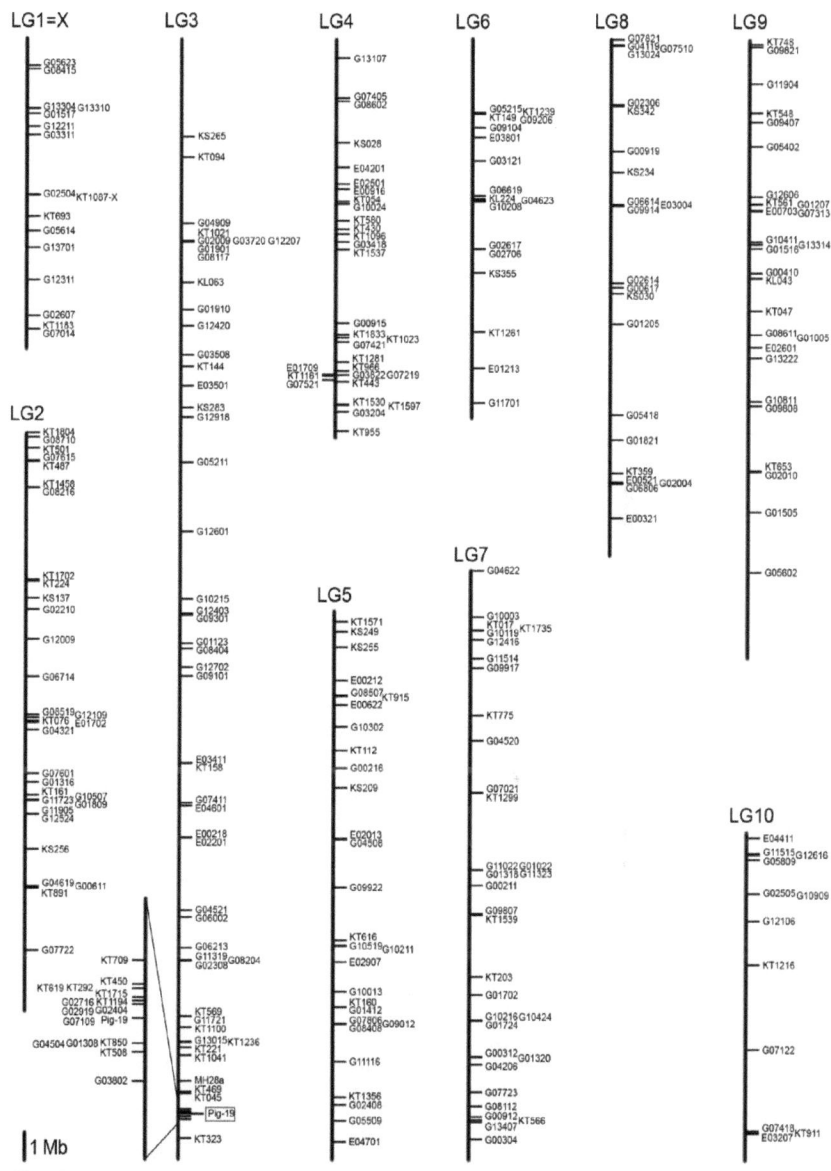

Figure 3

3.2.2 Insertion site analysis of lines generated in Göttingen

Not all details of my analysis are shown in the publication. Therefore I summarize the numbers found for the Göttingen lines here.

Within the GEKU insertional mutagenesis screen in Göttingen 2612 lines were fully analyzed for embryonic, larval, pupal and adult enhancer traps. 8.2 % (214) of the lines show enhancer traps and another 9.1 % (238) are homozygous lethal, 9 are female sterile. Of the lethal lines 13.9 % (33) additionally feature enhancer traps. The 238 lethal insertions are divided into 210 autosomal and 28 gonosomal insertions. The latter are all X-linked. The integration site of the donor construct was determined for all the lethal and sterile lines. Detailed analysis of the insertion sites demonstrates that the sterile insertions as well as the lethal insertions are mainly located close to or within gene predictions. Of the sterile insertions, 83% are located in introns or exons of GLEAN predictions (Table 1).

Table 1: Detailed analysis of sterile *piggy*Bac insertion sites

Sterile	n	%
Intron	3	50
CDS*	2	33
< 500 bp**	0	0
500 bp - 2500 bp**	0	0
distant (> 2500 bp)**	1	17
∑ localized	6	100
sequencing or blast problems	2	
not sequenced	1	
∑	9	

*exons, excluding UTRs
** distance to next gene prediction

The lethal lines show insertions of the *piggy*Bac element within exons, introns or closer than 500 bp to a GLEAN gene prediction in 79% (Table 2). Lethal or sterile insertions more distant to gene predictions are sparse. The real number of insertions close to genes is likely to be higher because most likely not all genes have been predicted. Hence lethal or sterile phenotypes are mainly due to a disruption of gene function caused by the integration of *piggy*Bac within the gene and not in regulatory regions.

Table 2: Detailed analysis of lethal piggyBac insertion sites

Lethal insertions	n	%
Intron	105	55
CDS*	26	14
< 500 bp**	19	10
500 bp - 2500 bp**	21	11
distant (> 2500 bp)**	20	10
∑ localized	191	100
sequencing or blast problems	42	
not sequenced	5	
∑	238	

*exons, excluding UTRs
** distance to next gene prediction

The integration site of 191 lethal and 6 viable lines could be determined and attributed to a chromosome. As the *Tribolium* genome has a size of about 200 Mbp about one insertion per Mbp should be achieved on every chromosome if there is no integration preference for any chromosome. The 197 localized piggyBac donor construct insertions exhibit a distribution from 0.7 to 1.3 insertions per Mbp (Tabelle 3). This is quite close to the expected 1.0 insertion per Mbp. This demonstrates that there is no strong piggyBac integration preference for any chromosome.

Table 3: Chromosomal location of piggyBac insertions

chromosome	n	%	chromosome size* [Mbp]	Insertions / Mbp
X	14	7	10,9	1,3
2	18	9	20,2	0,9
3	39	20	39	1,0
4	11	6	13,9	0,8
5	16	8	19,1	0,8
6	11	6	13,2	0,8
7	26	13	20,5	1,3
8	15	8	18	0,8
9	20	10	21,5	0,9
10	8	4	11,4	0,7
unmapped	19	10	12,3	
∑	197	100	200	

*based on NCBI map viewer, build 2.1

3.2.3 Rescreen of selected lethal lines

Cuticle preps of all 238 lethal lines generated in the insertional mutagenesis screen in Göttingen and additional 90 lethal lines from Kansas State University were analyzed for phenotypes. I mainly focused on head phenotypes but all other defects were also documented. The integration site and the probably affected gene were determined. These data were entered in the GEKU database (freely available at *www.geku-base.uni-goettingen.de*).

I further analyzed seven lines that appeared promising in the first screening round as head defects occured. The data gathered by these experiments are meant to confirm the findings of the screen and to provide further information for the selection of lines for future detailed analysis. For all these lines I first confirmed the integration sites via iPCR and determined the gene prediction(s) that were most likely hit by the insertion. I determined the *Drosophila* ortholog of the probably affected gene by blasting the predicted *Tribolium* protein sequence against the *Drosophila* genome at http://blast.ncbi.nlm.nih.gov/Blast.cgi. By backblast of the received *Drosophila* gene against the *Tribolium* genome at http://beetlebase.org/blast/blast.html I confirmed the orthology. Then I cloned the probably affected genes from cDNA with gene specific primers. With these genes RNAi experiments were performed and the phenotypes obtained by RNAi were compared to those of the respective lethal line. To that end I analyzed ten head cuticles of L1 larvae, if present, which correspond to 20 head halves from RNAi and insertion line as minor differences from one to the other half were observed. If the RNAi and mutant phenotypes were very similar I assumed that the cloned gene corresponds to the affected gene in the lethal line. In addition the expression pattern of the cloned gene was analyzed via whole-mount *in-situ* hybridization. The genomic flanking sequences of the *piggy*Bac donor construct integration sites as well as the sequences of the cloned genes, which were used as templates for antisense RNA probes and dsRNA synthesis, are shown in the appendix. Also the head bristle defects are summarized in table A1 in the appendix.

3.2.3.1 Analysis of lethal line G02408

In the first pass screen, several cuticles showed head bristle defects (see detailed analysis below). The *piggy*Bac donor construct is integrated on chromosome 5 at position 1890104 within the second intron of GLEAN_10926. This GLEAN is the ortholog of the *Drosophila* gene *CG6197/XAB2*. Hardly anything is known about the function of this gene. It is involved in DNA repair but not in DNA replication (Somma et al. 2008). The *Tribolium* cuticle phenotype is characterized by bristle pattern defects of the dorso-lateral larval head. Mainly affected are the bells of the bell row. They are distorted or partially missing in 70% and completely absent in 10% of the cuticles (n=20). The bell row bristle is also missing in 20%. The bristles of the vertex triplet are affected as well. The posterior vertex bristle is absent in 45% of the cuticles. The other vertex bristles and the vertex setae are affected at low frequencies. The posterior and anterior gena triplet setae are missing in 30% and 25%, respectively. The gena triplet bristle is absent in 30% of L1 cuticles (Figure 1). 40% of the cuticles exhibit weak abdominal bristle pattern defects as well and 20% show leg defects with bent tarsae. In addition only 41% of all laid eggs developed any cuticle. All the other eggs died prior to cuticle formation.

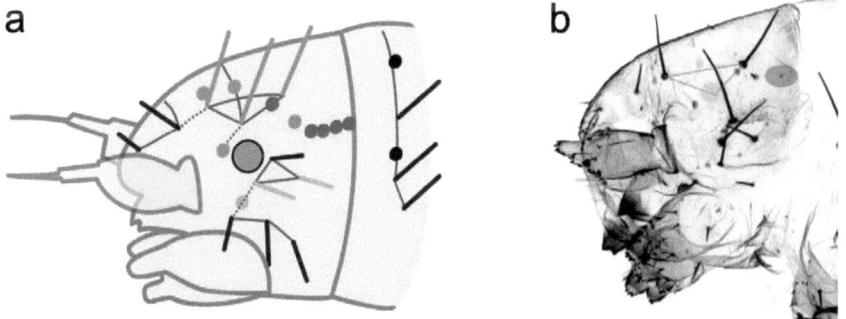

Figure 1: Mutant cuticle phenotype of lethal line G02408. (a) Schematic representation of a mutant head of a L1 larva cuticle. The frequency of absent or affected setae/bristles is indicated by different colors: Black: Wild type seta/bristle; Green: Distorted or absent in 6%-24% of cuticles; Orange: Distorted or absent in 25%-49% of cuticles; Red: Distorted or absent in 50%-100% of cuticles. (b) Head cuticle of mutant larva. In this cuticle the bell row (red circle), the ventral vertex seta and the median vertex bristle (green circles) are affected. The vertex triplet is affected in relatively low frequencies, whereas the bells and parts of the gena triplet are absent more frequently.

As the *Drosophila* ortholog of the probably affected gene is *XAB2* I cloned *Tc-XAB2*. In order to knock down *Tc-XAB2* in embryos and compare the mutant phenotype to the knock down of this gene, dsRNA was injected into *Tribolium* pupae (Bucher et al. 2002). No embryos could be obtained as all pupae died prior to eclosure. Hence, the gene function is essential during pupal stages.

In-situ hybridization with *Tc-XAB2* antisense RNA probe was performed to analyze the expression pattern. Ubiquitous expression could be observed during all embryonic germ band stages (Figure 2). This was not unexpected as XAB2 is involved in a basal cellular process.

Figure 2: Whole mount *in-situ* hybridization with *Tc-XAB2* antisense RNA probe (blue) and *Tc-wingless* antisense RNA probe (red). *Tc-XAB2* is expressed ubiquitously during all germ band stages of *Tribolium* embryonic development.

3.2.3.2 Analysis of lethal line G07411

This line was regarded as interesting in the screen due to the lack of the bell row. In this line the *piggy*Bac donor construct is integrated on chromosome 3 at position 12395367 within the third intron of GLEAN_03634. This gene is the ortholog of the *Drosophila* gene *CG10601/mirror* that encodes for a homeodomain transcription factor. The mutant phenotype is characterized by the loss of the bell row bells and bell row bristle in almost all cuticles (Figure 3). In some cases (37.5%; n=20) the posterior vertex bristle is absent as well. In addition the mutant phenotype exhibits additional setae at thoracic segments in 50% of cuticles. Only 15% of all laid eggs did not develop a cuticle which is the normal background. *Tc-mirror* has already been cloned and RNAi experiments have been performed (Posnien 2009). The phenotype caused by knock down via RNAi matches the mutant phenotype.

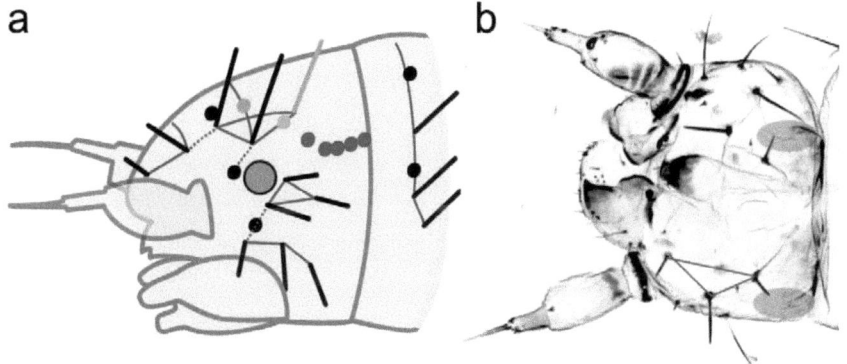

Figure 3: Cuticle phenotype of lethal line G07411. (a) Schematic representation of a mutant head of a L1 larva cuticle. The frequency of absent or affected setae/bristles is indicated by different colors: Black: Wild type seta/bristle; Green: Distorted or absent in 6%-24% of cuticles; Orange: Distorted or absent in 25%-49% of cuticles; Red: Distorted or absent in 50%-100% of cuticles. (b) Head cuticle of mutant larva. The bell row and the row bristle are affected in 100% of mutant cuticles (red circles).

Tc-mirror is expressed in an anterior median region during early germ band stages. Later on, the anterior expression expands laterally and also the stomodeal region expresses *Tc'mirror* as well as cells in the ventral head lobes in old germ band embryos (Posnien 2009).

3.2.3.3 Analysis of lethal line G07521

This line was noticed during the screening procedure as the distance between the ventral and anterior vertex setae seemed to be reduced. This line has probably two integrations of the *piggy*Bac donor construct. This was noticed as nearly all offspring of selected positively marked parents was also positively marked. In case of a single lethal insertion only 50% of the offspring should be positively marked with the dominant marker. One integration is within the first intron of GLEAN_00260. The orthologous *Drosophila* gene is *CG33531/Ddr* (*Discoidin domain receptor*). It encodes for a protein tyrosine kinase (Morrison et al. 2000) but no phenotypic data is available for *Drosophila*. The second insertion is 1.3 kb downstream of GLEAN_07654. The orthologous *Drosophila* gene is *CG33980/Vsx2/chx* (*visual system homeobox 2*). Vsx2 encodes for a transcription factor and is involved in development of the visual system. In *Drosophila* mutants of *Vsx1* and *Vsx2* the optic lobe and medulla are reduced in size. The latter may even be absent (Erclik et al. 2008).

The *Tribolium* mutant phenotype of line G07521 is characterized only by a reduced distance between the ventral and anterior vertex setae but the eyes are not affected (Figure 4). This phenotype appeared in relatively low frequencies (9%; n=32). Additionally, 55% of the embryos did not develop any cuticle. This suggests an important role of the affected gene during embryonic development prior to cuticle formation.

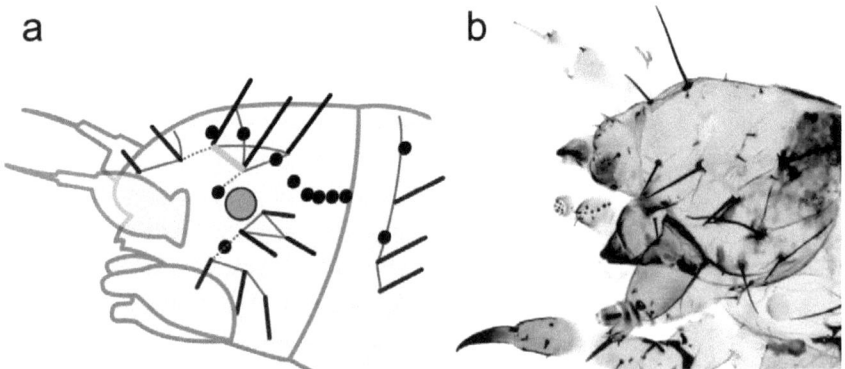

Figure 4: Mutant cuticle phenotype of line G07521. (a) Schematic representation of a mutant head of a L1 larva cuticle.. (b) Head cuticle of mutant larva. This mutation exhibits a reduced distance between the anterior and ventral vertex setae (marked by yellow line) in comparison to wild type L1 cuticles.

Tc'Vsx2 was knocked down via RNAi and knock down embryos have been analyzed for cuticle phenotypes. However, no cuticle did exhibit any phenotype. The eyes were not absent as well.

In-situ hybridizations were performed to analyze the expression pattern of *Tc-Vsx2*. It is expressed in an anterior median region from the stage on where also six *Tc-wingless* stripes are visible (arrow in **Figure 5** a, d). This expression domain remains stable throughout embryogenesis (**Figure 5** b, e). In late germ band stages an additional expression domain arises in the ocular region (**Figure 5** c, arrow in f).

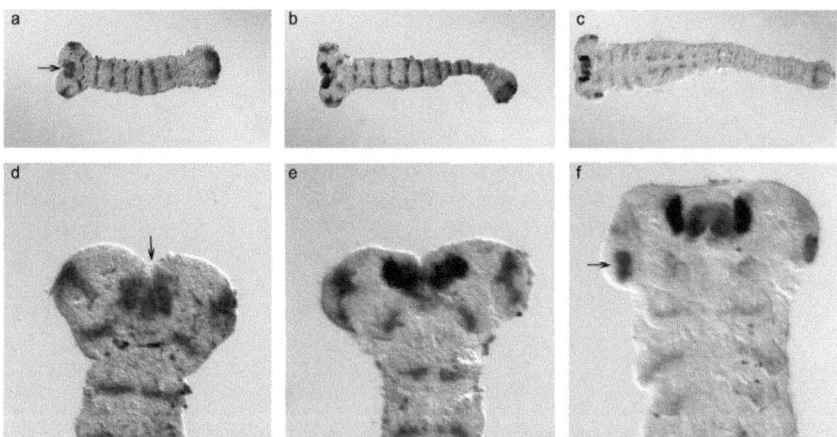

Figure 5: Whole mount *in-situ* hybridization with *Tc-Vsx2* antisense RNA probe (blue) and *Tc-wingless* antisense RNA probe (brown). (a, b) *Tc-Vsx2* is expressed in an anterior median region in elongating germ band stages of embryonic development. (c) Later on an additional expression domain arises in the ocular region. (d, e, f) Higher magnification of heads shown in (a, b, c).

3.2.3.4 Analysis of lethal line G09104

In this line the *piggy*Bac donor construct is integrated on chromosome 6 at position 10096530 within the third intron of GLEAN_14935. The orthologous *Drosophila* gene is *CG10260/phosphatidylinositol 4-kinase type 3 alpha* (*PI4KIIIα*). Phosphatidylinositol 4 kinases are involved in the metabolism of Phosphatidylinositol. They catalyze the conversion of Phosphatidylinositol (PI) to Phosphatidylinositol-(4)-phosphate. Phosphatidylinositolphosphates (PIPs) play an important role in many cellular processes like transcription and respond to morphogens and other signals (reviewed in (Skwarek and Boulianne 2009)).
No phenotypic data are available for knock down of *PI4KIIIα* in *Drosophila*. The *Tribolium* phenotype of the lethal line G09104 is characterized by a swollen head and a swollen first thoracic segment (Figure 6). This phenotype appeared only in very low frequencies (7%; n=27). Most cuticles were wild type but 40% of all laid eggs did not develop any cuticle. This indicates that this insertion causes a strong phenotype that leads to death during early embryonic development before cuticle secretion.

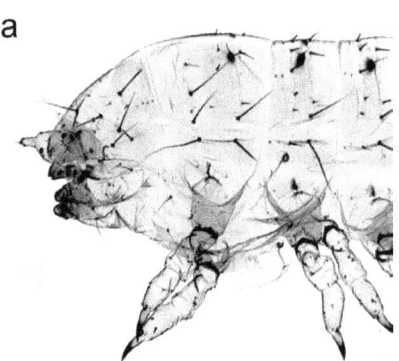

Figure 6: Mutant cuticle phenotype of line G09104. All appendages and head bristles are present but the head and the first thoracic segment are swollen.

The knock down via RNAi of *Tc-pi4kIIIα* leads to sterile beetles. This is not unexpected as PIPs are involved in many patterning processes.
Whole mount *in-situ* hybridization with a *Tc-pi4kIIIα* antisense RNA probe revealed ubiquitous expression of *Tc-pi4kIIIα* throughout all germ band stages of embryonic development (Figure 7).

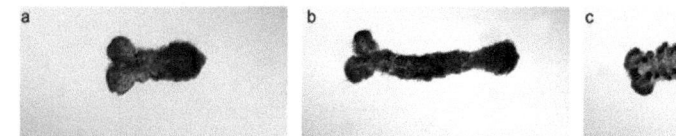

Figure 7: Whole mount *in-situ* hybridization with *Tc-Pi4KIIIα* antisense RNA probe (blue) and *Tc-wingless* antisense RNA probe (red). *Tc-Pi4KIIIα* is expressed ubiquitously during all germ band stages of *Tribolium* embryonic development.

3.2.3.5 Analysis of lethal line G10215

This line was noticed in the screen by distorted bristles of the gena triplet. The *piggy*Bac donor construct is integrated on chromosome 3 at position 19487977 within the first intron of GLEAN_00032. This gene is the ortholog of the *Drosophila* gene *CG8619/Cdc27*. *Cdc27* encodes a subunit of the anaphase promoting complex (APC/C). Larval neuroblasts mutant for *Cdc27* arrest in a metaphase-like state but some sister chromatids separate. Polyploidy appears in homozygous larval neuroblasts as well (Deak et al. 2003).

The *Tribolium* phenotype of the lethal line G10215 is characterized by the loss or distortion of several head bristles (Figure 8). Most of these defects occur at low frequency. The bell row bristle, antenna basal vertex bristle and median vertex bristle are absent in 7% (n=14) of cuticles but usually not all bristles are affected within the same cuticle. The anterior and posterior vertex triplet setae are absent in 21% and 14% and the triplet and anterior vertex bristles are missing in 14% of cuticles. The posterior and dorsal gena triplet bristles are missing in 21% and 14%, respectively. The bells of the bell row were distorted or partially absent most frequently (36%) but never completely lost. Trunk defects were constraint to weak bristle pattern defects in 43% of the mutant cuticles. Most of the laid eggs (81%) did develop a cuticle.

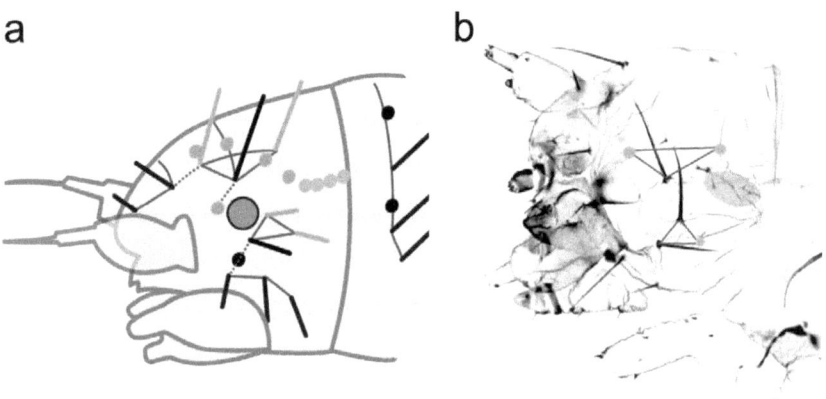

Figure 8: Cuticle phenotype of lethal line G10215. (a) Schematic representation of a mutant head of a L1 larva cuticle. The frequency of absent or affected setae/bristles is indicated by different colors: Black: Wild type seta/bristle; Green: Distorted or absent in 6%-24% of cuticles; Orange: Distorted or absent in 25%-49% of cuticles. (b) Head cuticle of mutant larva. In this cuticle the anterior and posterior vertex

Insertional mutagenesis screen: Rescreen of lethals-Results

setae, the posterior gena triplet (green circles) and the bell row bells (orange ellipse) are affected. Mainly affected were the bells of the bell row. The posterior and anterior vertex triplet setae and all bristles of the vertex triplet as well as the posterior and dorsal gena triplet are absent at low frequency.

Parental RNAi of *Tc-Cdc27* in pupae and adults leads to sterile beetles.

Whole mount *in-situ* hybridization with *Tc-Cdc27* antisense RNA probe revealed ubiquitous expression of *Tc-Cdc27* (Figure 9).

Figure 9: Whole mount *in-situ* hybridization with *Tc-Cdc27* antisense RNA probe (blue) and *Tc-wingless* antisense RNA probe (red). *Tc-Cdc27* is expressed ubiquitously during all germ band stages of *Tribolium* embryonic development.

3.2.3.6 Analysis of lethal line KT1269

In this line the *piggy*Bac donor construct is integrated within the first intron of GLEAN_13099. This gene is the ortholog of the *Drosophila* gene *CG4807/abrupt*. *Dm-abrupt* encodes a BTB zinc finger protein and is apart from other functions involved in dendrite morphogenesis, establishment and maintenance of neuromuscular junctions and also appendage morphogenesis (Hu et al. 1995; Sugimura et al. 2004).

Further sequence comparisons of GLEAN_13099 and *Dm-abrupt* revealed that also the adjacent gene prediction GLEAN_13098 belongs to the same gene. Hence *Tc-abrupt* has been predicted as two genes (GELAN_13098 and GLEAN_13099). The NCBI prediction (LOC663820) has correctly predicted *abrupt*. The NCBI prediction served as template for cloning *Tc-abrupt*.

The phenotype caused by this mutation is characterized by several defects. All mutant cuticles feature the loss of larval eyes and exhibit an additional tracheal opening at the third thoracic segment while wild type larvae possess tracheal openings only at the second thoracic segment and at all abdominal segments. The trochanter is missing at all leg pairs in more than 80% (n=12) of cuticles as well. The head bristle pattern exhibits severe defects. All vertex triplet setae and bristles are absent in 50% or more of the cuticles with exception of the anterior vertex bristle which is absent in 25%. The dorsal gena triplet and the gena triplet bristle are affected with high frequency as well (58% and 83%). The anterior and posterior maxilla escort are absent in 25% whereas the median maxilla escort is unaffected (Figure 10). In addition the labrum is sometimes malformed and has a bowl-like shape. This is probably a secondary affect as the tissue posterior to the labrum is distorted. Apparently, the connection between labrum and head is disordered.

The *Tc-abrupt* RNAi phenotype is characterized by the same features as the mutant phenotype. The main difference to the mutant phenotype is that the labrum is more often malformed and in addition to the other defects, the pygopods are shriveled. The frequency of head bristle defects is increased in the RNAi knock down cuticles compared to the mutant cuticles. In addition to the affected bristles in the mutant line also the labrum bristles (5%), the posterior and anterior gena triplet (10% and 45%) and the median maxilla escort (15%) are affected. Because of these additional effects I checked for off target effects that can occur if the dsRNA employed for RNAi has at least 21 bp sequence identity to another genomic region that could encode for a gene (Filipowicz 2005). Hence I blasted the sequence that was used as template for dsRNA synthesis against the *Tribolium* genome (Blast server of http://beetlebase.org/blast/blast.html). No nucleotide stretch of at least 21 bp with 100% identity was found outside of *Tc-abrupt*.

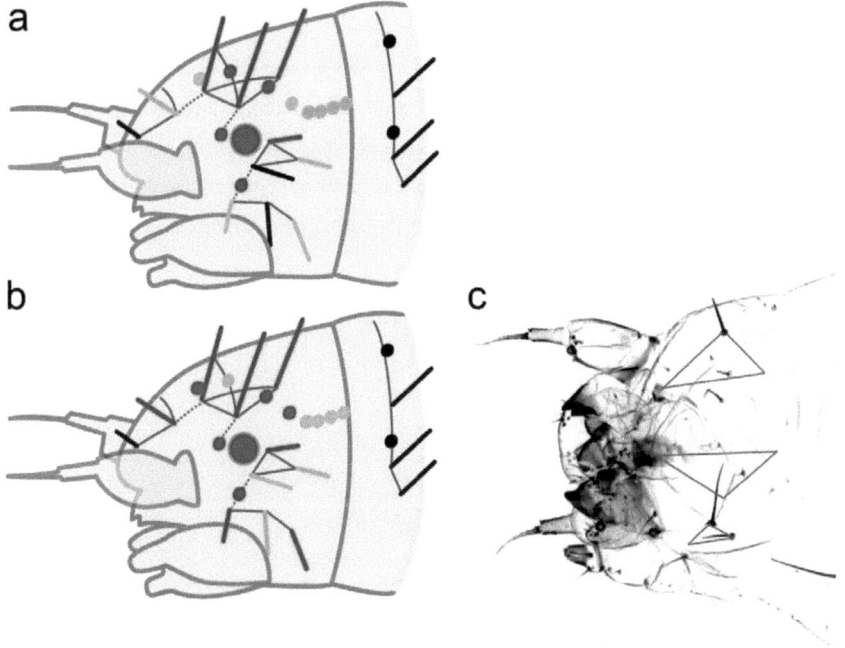

Figure 10: Cuticle phenotype of lethal line KT1269 and *Tc-abrupt* RNAi larvae. (a) Schematic representation of a mutant head of a L1 larva cuticle. (b) Schematic representation of *Tc-abrupt* RNAi knock down head of L1 larva cuticle. The frequency of absent or affected setae/bristles is indicated by different colors: Black: Wild type seta/bristle; Green: Distorted or absent in 6%-24% of cuticles; Orange: Distorted or absent in 25%-49% of cuticles; Red: Distorted or absent in 50%-100% of cuticles. (c) Head cuticle of *Tc-abrupt* RNAi L1 larva. The mutant and RNAi cuticles are characterized by the loss of the eyes, all bristles of the vertex triplet and the posterior gena triplet with high frequencies. The RNAi cuticles exhibit a stronger phenotype as bristles are absent with a higher frequency and more bristles are absent.

The expression pattern of *Tc-abrupt* was analyzed with a *Tc-abrupt* antisense RNA probe. First expression is visible in the posterior half of an early germ band stage embryo when *Tc-wingless* is expressed in an ocular stripe and posterior domain (Figure

11 a, d). Later on an anterior expression domain arises in anterior median tissue. This domain also expands laterally posterior to the ocular *Tc-wingless* domain. Additional broad and fuzzy segmental expression domains arise and the growth zone is *Tc-abrupt* positive as well (Figure 11 b, e). The anterior median expression domain expands to the anterior-most tissue of the embryo (Figure 11 c, f). These expression domains remain active during germ band elongation (Figure 11 g, k). In elongated germ band embryos the anterior median expression is restricted to the labrum anlagen. The stomodeal region is free of expression which leads to a separation of the anterior and the lateral posterior expression domain. The segmental expression is mainly restricted to the lateral parts of the thoracic and gnathal segments and the limb buds are *Tc-abrupt* positive (Figure 11 h, l). In retracting germ band stages, embryos exhibit strongest expression in the labrum, the limb buds of the gnathal segments and in the developing legs. Additionally *Tc-abrupt* is expressed in the abdomen in a lateral segmental manner (Figure 11 i, m).

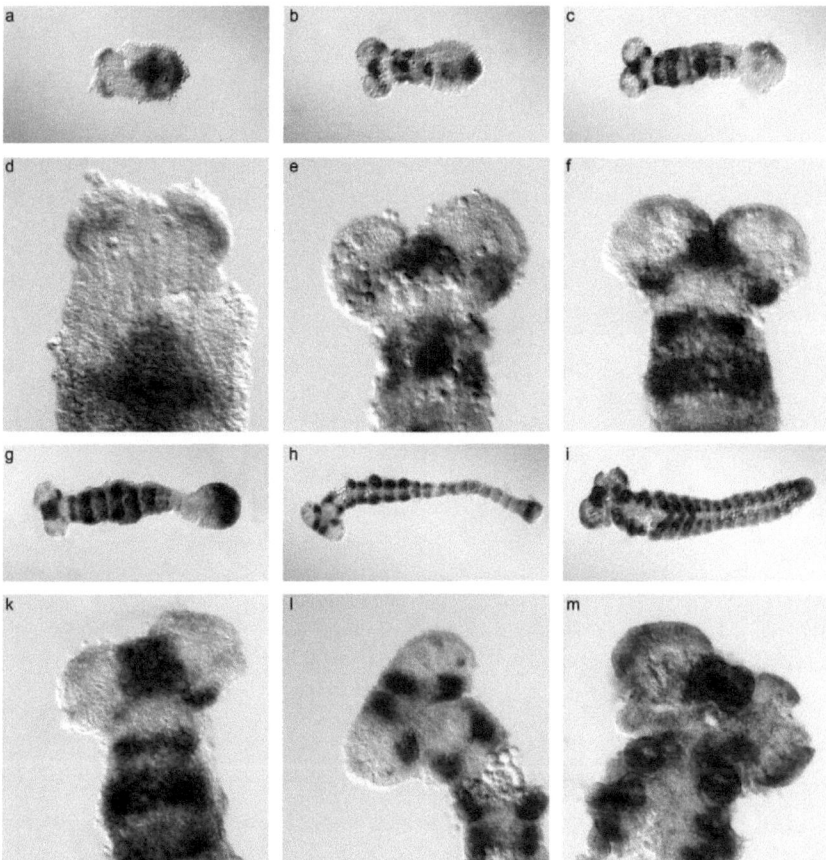

Figure 11: Whole mount *in-situ* hybridization with *Tc-abrupt* antisense RNA probe (blue) and *Tc-wingless* antisense RNA probe (red). (d,e,f,k,l,m) higher magnification of heads shown in (a,b,c,g,h,i). (a, d) *Tc-abrupt* expression starts in the posterior part of early germ band embryos. (b, e) During elongation an anterior median expression arises and expands laterally posterior to the ocular *wingless* domain. (c, f) Later on, the anterior expression expands anterior laterally into the antennal segment. (g, k) This expression remains active during elongation and (h, l) in fully elongated embryos the anterior median expression is restricted to the labrum whereas the stomodeum becomes free of expression. The domain posterior to the *wingless* domain stays active. Also all limb buds become *Tc-abrupt* positive whereas the segmental expression becomes weaker. (i, m) In retracted germ bands *Tc-abrupt* is mainly expressed in the labrum and all developing appendages. Also lateral segmental expression in the abdomen is visible.

3.2.3.7 Analysis of lethal line KS0294

This line was noticed in the screen due to the absence of the labrum. The *piggy*Bac donor construct is integrated within an unmapped stretch of genomic DNA for which no GLEAN gene predictions exist. Further analysis of a 20 kb region around the integration site with the prediction software AUGUSTUS revealed no ORFs of an appropriate size (Figure 12; Mario Stanke, pers. comm.). Thus this integration may be within a regulatory region or destroy other types of functional DNA elements.

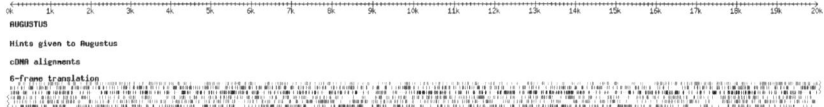

Figure 12: All possible ORFs 20 kb around the integration site. No single ORF of an appropriate size exists 10 kb upstream and 10 kb downstream of the integration site (Mario Stanke, pers. comm.).

The cuticle phenotype of homozygous L1 larvae for this mutation is characterized by the loss of the labrum. Also the dorsal anterior head tissue is affected. This is shown by the loss of bristles that mark this region (Figure 13). Mainly affected are the anterior vertex seta and anterior vertex bristle that are absent in 50% of cuticles (n=16). The more lateral head cuticle is hardly affected. The bells of the bell row are partially absent only in 6% of cuticles. The ventral vertex setae and antenna basal vertex bristles are lost in only 6% as well.

Insertional mutagenesis screen: Rescreen of lethals-Results

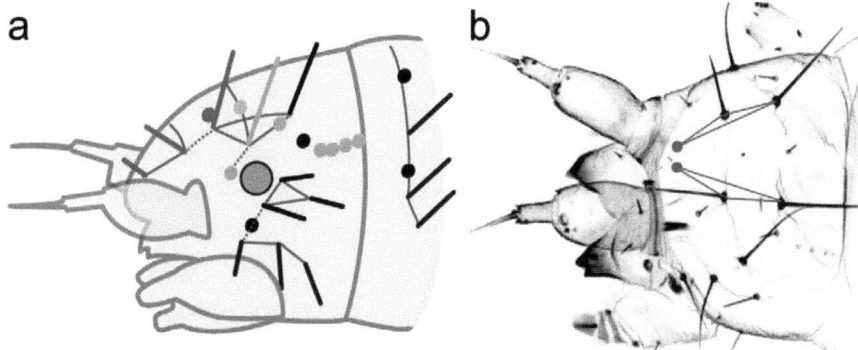

Figure 13: Cuticle phenotype of lethal line KS0294. (a) Schematic representation of a mutant head of a L1 larva cuticle. The frequency of absent or affected setae/bristles is indicated by different colors: Black: Wild type seta/bristle; Green: Distorted or absent in 6%-24% of cuticles; Orange: Distorted or absent in 25%-49% of cuticles; Red: Distorted or absent in 50%-100% of cuticles. (b) Head cuticle of mutant larva. The mutant cuticles are characterized by the loss of the labrum and anterior dorsal bristles (red circles).

No further analysis like RNAi and whole mount *in-situ* hybridization could be performed as it is unclear if this phenotype is caused directly by the loss of a gene function or is due to other reasons. It is also not known which gene could be affected to cause this phenotype.

3.2.3.8 Discussion of lethal screen

The integration sites of the *piggyBac* donor construct could be determined for all seven lines. One of the lines has two integrations but probably only one insertion is cause for the homozygous lethal phenotype. This was distinguished by the amount of positively marked offspring. The two different insertion sites could be determined via inverse PCR. One integration is within an intron of *Tc-Ddr*, the other one 1.3 kb downstream of *Tc-Vsx2*. As *Vsx2* is known to play a role in visual system development in *Drosophila* (Erclik et al. 2008), the *Tribolium* ortholog was thought to be a good candidate and was chosen for RNAi experiments. However, no cuticle phenotype could be observed. Hence, the RNAi experiments have to be repeated to confirm the absence of the transcript. If so, *Tc-Ddr* is probably the affected gene in the lethal line and should be analyzed by RNAi knock down. To get rid of the second insertion site the line will be outcrossed to wild type beetles in single pair matings. If not more than 50% of the offspring is positively marked, their offspring will be checked for the cuticle phenotype. If about 25% of the offspring exhibits the mutant phenotype, the single pair mating can be used to establish a new stock with only one integration site.

The lethal lines G09104, G10215 and G02408 exhibit head phenotypes. The insertion sites have been determined and the probably affected genes were cloned. However, parental RNAi experiments (Bucher et al. 2002) could not confirm the mutant cuticle phenotype as the knock down of the respective genes lead to sterility of the adults or death of the injected pupae prior to eclosure, respectively. To circumvent these problems caused by parental RNAi the dsRNA of the genes of interest can be injected into embryos of different stages. Thus, early to late function of the respective genes can be analyzed and the phenotypes can be compared to the mutant ones. The phenotype of line G09104 is especially interesting as the enlarged head and first thoracic segment could be connected with allometric growth. To determine where the tissue is expanded (dorsal, ventral, ubiquitously, etc.) morphometric data are required. The rescreen of line G10215 revealed that the phenotype is not very interesting. This is due to the low penetrance and the high variability of the bristle defects. This is not unexpected as the probably affected gene *Tc-Cdc27* is most likely a general cellular factor. The high number of eggs that did not develop any cuticle in line G02408 indicates that the cuticle phenotype represents a hypomorphic situation because many severe patterning defects interfere with cuticle formation. The involvement of the probably affected gene *Tc-XAB2* in a basal cellular process and its ubiquitous expression argue against a direct role in patterning. Therefore this line may be not so interesting.

The lethal line KS0294 exhibits an utmost interesting and specific phenotype with the complete loss of the labrum. A similar phenotype has only been known from the knock down of *Tc-optix/six3* and *Tc-cap-n-collar*. The integration site was determined but no gene prediction is found close to the insertion site. No open reading frame of a reasonable size could be found 20 kb around the insertion site (Mario Stanke, pers. comm.). Hence, the insertion of the *piggyBac* donor construct has probably disrupted other functional DNA elements or is lying in a very large intron. Because the insertion is located on the unmapped part of the genomic sequence it could theoretically affect *Tc-optix/six3* or *Tc-cap-n-collar*. Alternatively, transcription factor binding sites of a regulatory region may be affected. Another possibility is that this region encodes for small noncoding regulatory RNAs. These RNAs have been suggested to play an important role in gene regulation from morphologically simple to complex animals (Grimson et al. 2008). Further analysis of the genomic region is necessary to understand the direct cause of this very interesting phenotype.

Tc-mirror is the affected gene in line G07411 as the mutant phenotype was confirmed by RNAi (Posnien 2009). In *Drosophila*, it is not known that *mirror* plays a role in head patterning. It is only involved in forming the equator in the compound eye (McNeill et al. 1997) but the vertebrate ortholog *irx* is important for neural plate patterning (Briscoe et al. 2000). The relationship between expression pattern and knock down phenotype is not trivial as the anterior median region where *Tc-mirror* is expressed gives rise to the labrum but the lateral head cuticle is affected in knock down cuticles. Nevertheless this mutation causes a very specific and interesting phenotype that should be followed up.

Tc-abrupt is the mutated gene in line KT1269. This was confirmed by knock down of *Tc-abrupt* via RNAi and the phenotype occurred with high penetrance. From *Drosophila*, it is known that *abrupt* influences neuro-muscular connections and the branching of neurons is affected in mutants. Mechanosensory bristles on the thorax and wings are missing and in addition the distal parts of the legs are malformed or absent in ab[1]/ab[1D] heterozygous adults (Hu et al. 1995; Li et al. 2004). These phenotypes are similar to those monitored in *Tc-abrupt* mutant or knock down *Tribolium* larvae. In contrast to *Drosophila* in *Tribolium* the head bristles are affected and not thoracic ones.

Also no head phenotype is reported in *Drosophila*. It was shown that the *Tc-abrupt* RNAi phenotype is stronger than the mutant one and in addition affects the pygopods. Due to the integration of the donor construct within an intron of *Tc-abrupt* the transcript could be spliced in the correct manner in rare cases. Hence, these transcripts could rescue the mutant phenotype to some extent. The pygopods phenotype in RNAi may indicate a lower sensitivity of this tissue to *Tc-abrupt* reduction. All together this is a very interesting phenotype and should be analyzed with priority.

Taken together, four of the seven lines that were analyzed in the re-screen turn out to be interesting. As I have identified about 40 interesting lines in the first pass screen, I would expect that at least another 20 may prove to be revealing. Importantly, this work shows that genes not known to be involved in head development can be identified by this screen.

3.2.4 Rescreen of selected enhancer trap lines

The 2612 viable lines generated in Göttingen were screened for embryonic, larval pupal and adult enhancer traps. 214 lines exhibit an enhancer trap in at least one of these stages. The screen for enhancer traps was conducted by checking for EGFP fluorescence in tissue additional to the eyes. The fluorescence in the eyes is due to the transformation marker 3xP3-EGFP that drives expression of EGFP in larval, pupal and adult eyes (Berghammer et al. 1999). Out of the 214 lines several were selected for further analysis because of their interesting patterns. This analysis included the determination of the integration site by iPCR and *in-situ* hybridization with an EGFP antisense RNA probe. In addition to the *Tc-Ultrabithorax (Tc-Ubx)* and *Tc-lame duck (Tc-lmd)* enhancer trap lines shown in the manuscript (see chapter 3.2.1) the following three selected lines exhibit fluorescence of EGFP in different kinds of tissues.

3.2.4.1 Enhancer trap line G03920

This line was noticed in the screen as fluorescence of EGFP was detected in the stomodeal region and the foregut. The *piggy*Bac donor construct is integrated on chromosome 7 at position 13567799. This integration is 13.4 kb downstream of the next GLEAN gene prediction and within an intron of a gene predicted by NCBI. The *Drosophila* ortholog of GLEAN_08936 is *CG3090/Sox-14*. *Sox-14* is expressed ubiquitously in *Drosophila* embryos and knock down leads to salivary gland destruction and defects in the larval midgut (Chittaranjan et al. 2009). The *Drosophila* ortholog of the gene predicted by NCBI LOC661337 is *CG11084/prickle*. *prickle* is strongly expressed in cells that are involved in morphogenetic movements, like cells in the cephalic furrow and dorsal fold. In later stages it is also expressed in the parasegmental folds (Gubb et al. 1999).

The enhancer trap line shows strong fluorescence of EGFP in the stomodeal region and the foregut (Figure 14 a). In an older embryo fluorescence is visible in a lateral segmental pattern from a dorsal view (Figure 14 b). *In-situ* hybridization with an EGFP antisense RNA probe within the enhancer trap line revealed expression in the stomodeum and proctodeum (Figure 14 c, d).

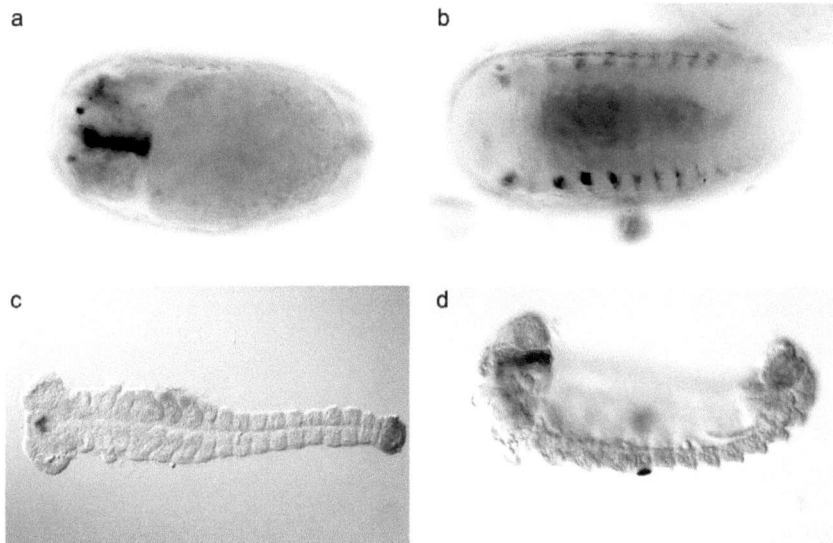

Figure 14: Fluorescence and expression of EGFP in enhancer trap line G03920. Pictures (a) and (b) have been inverted such that the fluorescence is shown in black. Fluorescence of EGFP (a) in the stomodeal region in a ventral view; (b) in a segmental pattern in the thorax and abdomen in a dorsal view. (c, d) Expression of *EGFP* within the enhancer trap line revealed by *in-situ* hybridization with EGFP antisense RNA probe. Expression is seen in: (c) the stomodeum and proctodeum of retracting embryos; ventral view; (d) the stomodeal region and the proctodeum in embryos close to dorsal closure; dorso-lateral view.

3.2.4.2 Enhancer trap line G10011

This line was noticed in the screen by fluorescence in the lateral head lobes. The piggyBac donor construct has integrated on chromosome 4 at position 6843201. This integration is within introns of two different gene predictions (TC008169 and NCBI prediction LOC664429) (Figure 15). The Drosophila ortholog of TC008169 is *CG17271*. The Drosophila ortholog of LOC664429 is *CG32532* and encodes for a homeodomain transcription factor. No information is available for either of these genes.

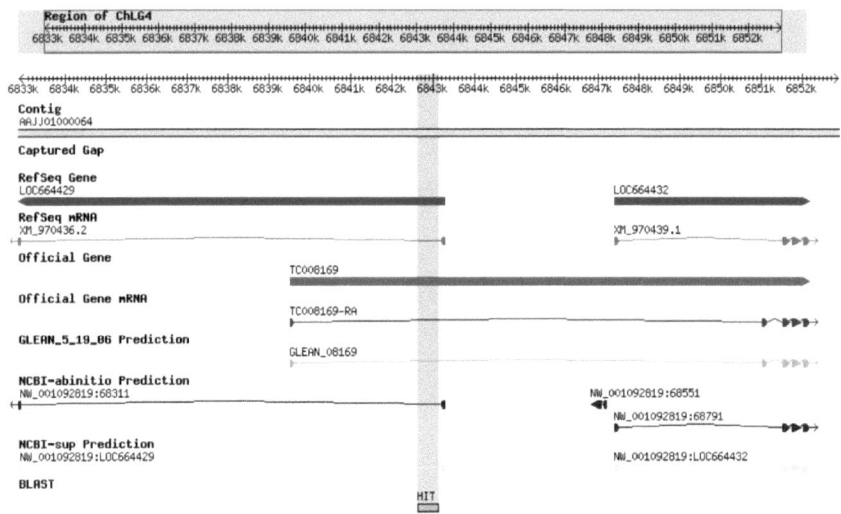

Figure 15: Blast result of flanking sequence of line G10011 blasted against *Tribolium* genome at http://beetlebase.org/. The piggyBac donor construct is integrated between the gene predictions TC008169 and LOC664429.

The embryos shown in Figure 16 are in a similar developmental stage. The enhancer trap line exhibits fluorescence in the lateral head lobes (arrow in Figure 16 a) and expression of EGFP was also detected in this region by *in-situ* hybridization (arrow in Figure 16 b).

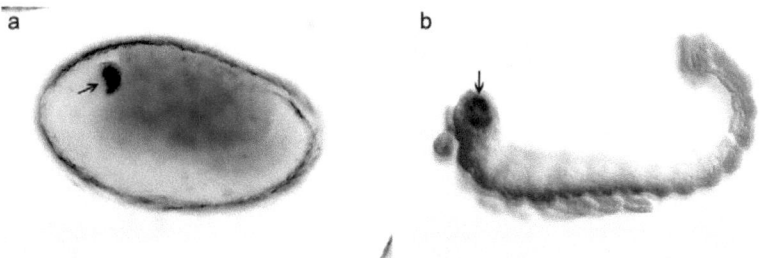

Figure 16: Fluorescence and expression of EGFP in enhancer trap line G10011. Picture (a) has been inverted such that the fluorescence is shown in black. (a) lateral view; Fluorescence of EGFP in the lateral head lobe of a living embryo. (b) lateral view; Expression of EGFP within the enhancer trap line revealed by *in-situ* hybridization with EGFP antisense RNA probe. Expression is seen in the lateral head lobe.

3.2.4.3   Enhancer trap line G11122

This line was noticed in the screen by EGFP fluorescence in an anterior median region of embryos. The piggyBac donor construct has integrated on chromosome 3 at position 2296104. The integration is within an intron of GLEAN_03098. The Drosophila ortholog of this gene is CG12052/lola. Drosophila lola is expressed in the central nervous system as well as in the peripheral nervous system in embryonic stages. Expression was also detected in the optic lobe placode (Giniger et al. 1994). The enhancer trap line exhibits fluorescence in the stomodeum (Figure 17 a). In-situ hybridization with an EGFP probe within this enhancer trap line revealed expression in the stomodeum and proctodeum (Figure 17 b, c). EGFP is also expressed in a spot in the ocular region (arrows in Figure 17 b).

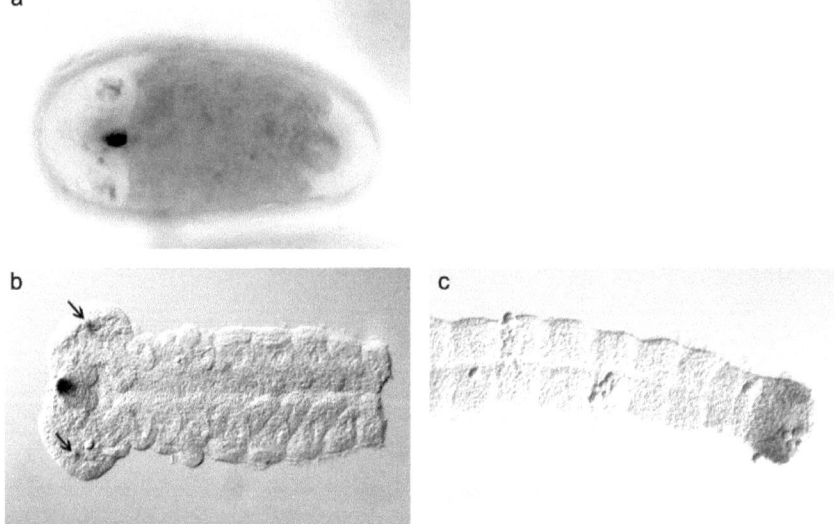

Figure 17: Fluorescence and expression of EGFP in enhancer trap line G11122. Picture (a) has been inverted such that the fluorescence is shown in black. (a) Ventral view; Fluorescence of EGFP in the anterior median region that probably corresponds to the stomodeum in a living embryo. (b, c) Ventral view; Expression of EGFP revealed by in-situ hybridization with EGFP antisense RNA probe. Expression is seen in the lateral head lobe. (b) EGFP is expressed in the stomodeum and in a spot in the ocular region (arrows). (c) EGFP is also expressed in the proctodeum.

3.2.4.4 Discussion of enhancer trap screen

The enhancer trap screen revealed fluorescence of EGFP in different kinds of tissues and within all stages of *Tribolium* development. 3xP3-EGFP which is part of the mutator construct serves as transformation marker as it drives expression of EGFP in the larval, pupal and adult eyes (Berghammer et al. 1999) but it also can pick up enhancer traps (Lorenzen et al. 2007). Indeed, many different patterns have been detected (see GEKU base at www.geku-base.uni-goettingen.de). However, quite often fluorescence was also observed in the brain as well as in the entire central nervous system. Apparently, this is due to the 3xP3 promoter that appears to have a bias towards CNS patterns and by itself generates a pattern in the brain. This was a drawback for the search for enhancer traps in the head as it is often not possible to distinguish between background fluorescence of the 3xP3 promoter driven EGFP expression and expression that is caused by an endogenous promoter. Nevertheless, several enhancer trap lines that show fluorescence of EGFP in the head were detected. This was possible if the enhancer trap caused early expression, because 3xP3 driven expression comes up rather late during embryonic development.

Enhancers can act over large distances. Thus, the identification of the gene that exhibits the same expression as EGFP in the enhancer trap line is quite difficult as the integration may be in large distance to the respective gene.

In the enhancer trap line G03920 the *piggy*Bac donor construct is integrated within an intron of the *Tribolium* ortholog of *prickle*. This line exhibits strong fluorescence of EGFP in the stomodeum and proctodeum as well as in a segmental manner in old embryos. Only the segmental expression may correspond to the expression seen in *Drosophila* within the parasegmental folds (Gubb et al. 1999). However, also formation of the stomodeum may involve morphogenetic movements which would fit to *prickle* function. The expression domains in the stomodeum and proctodeum do most likely not reflect activity of the second closest gene prediction *Sox-14*, as *Sox-14* is expressed ubiquitously in *Drosophila*. Hence, this expression may be due to the activation by other enhancer elements and/or the combination of several ones that also interact with the *piggy*Bac donor construct. The stomodeal pattern might be useful with respect to the development of the brain that develops directly adjacent and the stomatogastric nervous system that arises from the stomodeum roof.

In enhancer trap line G11122 a similar expression pattern as in line G03920 within the stomodeum and proctodeum is seen, but the expression in the stomodeum is more defined in line G11122. EGFP is also expressed in a spot in the ocular region. This may be expression in the optic lobe placode as this is also seen for *Drosophila lola*. In *Drosophila* additional expression domains in the central and peripheral nervous system are described (Giniger et al. 1994) that were not detected in *Tribolium*. Nevertheless, the expression domain in the ocular region is distinct and quite interesting. It could serve to follow the morhogenetic movements and in addition could mark a subset of neuroblasts and their offspring. It is possible that the 3xP3 promoter of the *piggy*Bac donor construct only interacts with a fraction of enhancer elements and thus leads to a partial expression of EGFP in the enhancer trap line.

The similarity of expression in two lines with different insertion sites might also indicate that the 3xP3 based reporter is somehow biased to this type of pattern. Indeed, it is known that besides the expected Pax6 driven pattern also additional patterns arise in the brain (Koniszewski, pers. comm.).

The enhancer trap line G10011 exhibits a very interesting fluorescence of EGFP within the lateral head lobes. The *piggy*Bac donor construct has integrated within introns of two differentially predicted genes. As information is available for neither of the predicted genes, it cannot be assumed which of the genes is probably expressed in the respective domain. This line may be useful to image the morhogenetic movements involved in head formation.

In order to identify genes that are most likely under the control of the same enhancer elements as the *piggy*Bac donor construct, the candidates have to be cloned and *in-situ* hybridization with a gene specific RNA antisense probe has to be performed within wild type embryos. By this the gene specific expression pattern can be compared to the expression of EGFP. If the gene specific expression pattern is the same or at least similar to the expression of EGFP in the respective enhancer trap line, the gene is most probably identified. In case of the line G10011 this might reveal a novel gene involved in head development.

In summary, although many different partterns have been identified, the construct is not optimal for detecting enhancer traps in the head. First, its own activity in the head obscures interesting novel patterns. Second, it appears to have a bias for certain patterns. An enhanced enhancer trap screen would involve a marker not based on EGFP but on the *vermillion* gene that rescues white eyes. Moreover, it should contain an endogenous promoter and it could be combined with GAL4 instead of EGFP. This would allow to misexpress genes in the respective patterns.

3.3 The binary expression system GAL4/UAS in *Tribolium*

In this part, the establishment of the binary expression system GAL4/UAS is described and its properties further characterized. I also found that basal promoters from *Drosophila* and artificial promoters are much less efficient compared to endogenous promoters. The GAL4/UAS system will be a useful tool to unravel gene function in general and for the analysis of head development specifically.

Schinko JB, Weber M., Viktorinova I., Kiupakis A., Averof M, Klingler M., Wimmer EA, Bucher G.

Authors contributions to the practical work:

Schinko JB: All practical work except the ones listed below:

Weber M.: Construction and unsuccessful test of: pBac[3xP3-EYFP;UAS-Dm'hsp70TATA-LacZ]

Kiupakis A.: Construction of: pBac[3xP3-EGFP;gUAS-SCP1-DsRed]

Viktorinova, I.: Construction and unsuccessful test of: pBac[3xP3-Dsred;UAS-Dm'hsp70TATA-LacZ]; pBac[3xP3-EYFP;3xP3-GAL4Δ]; pBac[3xP3-EYFP;3xP3-GAL4-VP16]

Bucher G: Construction and unsuccessful test of pBac[3xP3-EGFP;Dm'hs-GAL4] and pBac[3xP3-DsRed;UAS-Dm'hsp70TATA-Tc'giant-SV40]

Status: **Submitted** to BMC Developmental Biology

Abstract

The red flour beetle *Tribolium castaneum* has developed into an insect model system second only to *Drosophila*. Moreover, as coleopteran it represents the most species rich metazoan taxon on earth.The genetic toolbox has expanded enormously in the past years but spatio-temporal misexpression of genes has not been possible so far. Binary expression systems allow analyzing gene function by spatially and temporally controlled misexpression. Here we report on the establishment of the GAL4/UAS system in *Tribolium castaneum*. Both GAL4Δ and GAL4VP16 driven by the endogenous heat shock inducible promoter of the *Tc-hsp68* gene are efficient in activating reporter gene expression under the control of the Upstream Activating Sequence (UAS). UAS driven ubiquitous tGFP fluorescence was observed in embryos within four hours after activation while *in-situ* hybridization against tGFP revealed the onset of expression already after two hours. This response is quick in relation to the embryonic development of *Tribolium* which takes 72 hours with segmentation being completed after 24 hours. This allows to make use of the system for the study of segmentation and other embryonic processes. Moreover, we find that the use of *Tribolium* basal promoters is essential for transgenic constructs while the direct use of *Drosophila* constructs in *Tribolium* is not recommended.

Introduction

The red flour beetle *Tribolium castaneum* has evolved into an important insect model system that – being a coleopteran - represents one fourth of all described animal species (Hunt et al. 2007) including many pest species (boll weevil, corn rootworm, Colorado potato beetle and Asian longhorn beetle). While the technical amenability of the fruit fly *Drosophila melanogaster* remains unmatched there are topics that for different reasons cannot be answered in the fly. Evolutionary questions require comparative functional data from other insects and many processes are derived in *Drosophila* and are, hence, not representative for insects. For instance, segments are specified all at one time in *Drosophila* (long germ mode) instead of sequential formation in most insects (short germ mode), embryonic legs are not developed in *Drosophila* while insect larvae usually do have functional appendages, extraembryonic membranes are highly reduced and the head becomes highly reduced and involuted during embryogenesis in *Drosophila*. Some issues of insect biology cannot be studied in *Drosophila* because it lacks the respective character. One example are the odoriferous defensive glands that play a crucial role in insect communication and defense but are not found in *Drosophila*. Hence, there is a need for complementary insect model systems for comparative functional work and for studying processes that for one reason or the other are hard to study in *Drosophila*.

The recent development of several techniques has rendered *Tribolium* the second best insect model system. Its genome is sequenced (Richards et al. 2008) and germ line transformation in *Tribolium* is as efficient as in *Drosophila* and several marker systems for following gene transfer are available (Berghammer et al. 1999; Horn et al. 2002; Lorenzen et al. 2003). Based on these systems, an insertional mutagenesis system has been established (Lorenzen et al. 2007) which has been used to generate a collection of enhancer trap and homozygous lethal lines (Trauner et al., submitted). Importantly, robust RNAi techniques are established. RNAi can be applied by embryonic injection but the systemic spread of the RNAi response also allows injection of females and analysis of the phenotype in the offspring. Injection of larvae leads to phenotypes during metamorphosis without affecting their early gene function (Brown et al. 1999; Bucher et al. 2002; Tomoyasu and Denell 2004; Konopova and Jindra 2007; Konopova and Jindra

2008; Miller et al. 2008; Tomoyasu et al. 2008; Suzuki et al. 2009). Apparently, all tissues can be targeted by RNAi (Miller et al. 2008) and the Null phenotype can be phenocopied by RNAi (Cerny et al. 2008).

While knock down of gene function via RNAi is extremely efficient in *Tribolium*, the spatio-temporally controlled misexpression of genes has not been possible. Binary expression systems have the advantage that any gene can – depending on the availability of driver lines - be expressed in tissue-specific patterns or at different developmental stages (Bello et al. 1998; Szuts and Bienz 2000; Stebbins et al. 2001; McGuire et al. 2004; Viktorinova and Wimmer 2007; Zhong and Yedvobnick 2009). This includes dominant lethal or sterility inducing genes because the gene is only activated when the driver and responder activities are combined. One widely used binary expression system is the GAL4/UAS system.

GAL4 was identified in the yeast *Saccharomyces cerevisiae* as a regulator of *GAL1, GAL10* and other genes induced by galactose (Laughon et al. 1984; Laughon and Gesteland 1984). GAL4 regulates transcription by binding to a 17 bp site in the Upstream Activating Sequence (UAS) (Giniger et al. 1985). The GAL4/UAS system consists of a driver construct, where expression of the heterologous transactivator GAL4 is driven by an inducible or tissue specific enhancer. In the responder construct, the gene of interest is under the control of the heterologous GAL4-controlled Upstream Activating Sequence (UAS) (Fischer et al. 1988; Brand and Perrimon 1993; Rorth 1998). For driver and responder, separate transgenic lines are generated that are viable. Upon crossing these strains, the gene of interest is expressed in the progeny in the pattern defined by the driver.

The UAS/Gal4 system has been established in *Drosophila* and has become a standard technique adapted to diverse uses. One of the numerous extensions of this system is for example GAL80 that binds to the carboxy-terminal amino acids of GAL4 and inhibits activation of transcription (Ma and Ptashne 1987; Fischer et al. 1988; Brand and Perrimon 1993; Rorth 1998; Duffy 2002; McGuire et al. 2004; Viktorinova and Wimmer 2007). Moreover, it has been adapted to mouse (Ornitz et al. 1991), zebrafish (Scheer and Campos-Ortega 1999), *Xenopus* (Hartley et al. 2002) and *Arabidopsis* (Guyer et al. 1998).

Further analysis of the GAL4 transactivator revealed that it consists of two functional domains. The DNA binding domain maps to the first 74 amino acids whereas the activation domain maps to two regions, amino acids 148–196 and 768–881. In the transactivator version GAL4Δ, the activation domain is directly fused to its DNA binding domain (Ma and Ptashne 1987). This results in a smaller and thus probably faster synthesized protein which results in a shorter lag between Gal4Δ expression and activation of target genes. It has also been shown that GAL4Δ activates reporter gene expression twice as effectively as the original GAL4 in *Drosophila* (Viktorinova and Wimmer 2007).

In the GAL4-VP16 version, the activation domain of GAL4 has been replaced by the highly acidic portion of the herpes simplex virus protein VP16 that activates transcription of immediate early viral genes (McKnight et al. 1987; O'Hare and Goding 1988; O'Hare et al. 1988; Preston et al. 1988; Triezenberg et al. 1988; Triezenberg et al. 1988). It has been shown that GAL4-VP16 can efficiently activate transcription in mammalian cells (Sadowski et al. 1988). Also in *Drosophila* GAL4-VP16 has a higher potential to activate reporter gene expression than GAL4, but it has shown to be less efficient compared to GAL4Δ (Viktorinova and Wimmer 2007).

With this work we establish the GAL4/UAS system in *Tribolium*. We show that both GAL4Δ and GAL4-VP16 transactivate well, with GAL4Δ being slightly more efficient. Moreover, we show that the use of *Tribolium* endogenous basal promoters is essential for efficient expression.

.

Material and Methods

Constructs

All transactivator and responder constructs were stably integrated into the genome by transposition using the *piggy*Bac vectors pBac[3xP3-EGFPafm], pBac[3xP3-ECFPaf] (Horn and Wimmer 2000), pBac[3xP3DsRedaf] (Horn et al. 2002) or pXL-BacII[3xP3-EYFPaf] (Schinko 2003). The sequence of those constructs that turned out to work are shown in the appendix. Maps of constructs are depicted in Figure 1. Detailed maps are available from the authors.

Transactivator plasmids

pBac[3xP3-ECFP;Tc'hsp5'-GAL4Δ-3'UTR]

pBac[3xP3-EGFP;Tc'hsp5'-GAL4Δ-3'UTR]

pBac[3xP3-EGFP;Tc'hsp5'-GAL4-VP16-3'UTR]

pBac[3xP3-EGFP;Dm'hs-GAL4] (provided by Gregor Bucher)

pBac[3xP3-EYFP;3xP3-GAL4Δ] (provided by Ernst Wimmer)

pBac[3xP3-EYFP;3xP3-GAL4-VP16] (provided by Ernst Wimmer)

Responder plasmids

pXL-BacII[3xP3-DsRed;UAS-Dm'hsp70TATA-Tc'h_p-EYFP]

pBac[3xP3-DsRed;UAS-Dm'hsp70TATA-Tc'giant-SV40] (provided by Gregor Bucher)

pBac[3xP3-EGFP;gUAS-SCP1-DsRed] (provided by Michalis Averof)

pBac[3xP3-Dsred;UAS-Dm'hsp70TATA-LacZ] (provided by Ernst Wimmer)

pBac[3xP3-DsRed;UAS-Tc'hsp68_p-tGFP]

Tribolium stocks and germline transformation

Tribolium germline transformation was performed according to standard procedure into preblastoderm embryos of the *vermillion white* (*vw*) strain (Berghammer et al. 1999; Lorenzen et al. 2003), by using *piggy*Bac constructs, at a concentration of 500 ng/µl in injection buffer (5 mM KCl, 0.1 mM KH_2PO_4, 0.1 mM Na_2HPO_4 pH 6.8) together with 300 ng/µl helper plasmid phspBac (Handler and Harrell 1999). Femto Jet (Eppendorf) device with pulled and cut borosilicate glass capillaries were used for injections. Injected embryos were kept under humid conditions for two days at 32°C/89.6°F, afterwards transferred to less humidity and kept until they hatched at 32°C/89.6°F. Larvae were collected and transferred to full wheat flour. Adult G0 beetles were crossed to *vw* wild type strain. Transgenic beetles were outcrossed with *vw* wild type and kept as homozygous or heterozygous stocks.

Transformation markers and epifluorescence microscopy

As transformation markers EGFP (Cormack et al. 1996; Yang et al. 1996) (Clontech Laboratories, Inc. Palo Alto, CA), EYFP (Cubitt et al. 1999), ECFP (Patterson et al. 2001) as well as humanized variant DsRed1 (Horn et al. 2002) were used. The 3xP3-driven expression pattern of the fluorescent markers was detected in the eyes of *T. castaneum* by the Leica MZ 16FA fluorescence stereomicroscope with planachromatic 0.8 x objective. Different filter sets were used: EGFP_LP, ECFP_LP and DsRedwide.

Tribolium crosses

To activate the binary expression system, adult beetles selected for the dominant markers of the transactivator and responder lines were crossed together and kept for 10 days at 28°C/82.4°F. Within this period of time beetles of the transactivator and responder lines mate and the sperm of previous matings with males of the same line will be largely replaced. Afterwards the crosses were transferred to fresh flour. 24 hours egg collections at 32°C/89.6°F or transheterozygous larvae, pupae and adults were heat shocked and analyzed for tGFP fluorescence. For whole mount *in-situ* hybridizations, matings were kept for 72 hours at 25°C/77°F for egg collection.

Heat shock conditions

Embryonic heat shocks were performed in 1.5 ml Eppendorf tubes in a water bath for 10 minutes at 46°C/114.8°F. Larval, pupal and adult heat shocks were performed in 2 ml Eppendorf tubes for 20 minutes at 46°C/114.8°F.

Detection of tGFP fluorescence

Eggs were dechorionated under mild conditions in Natriumhypochlorite (1% DanKlorix), and subsequently aligned on a microscope slide. These steps were performed at RT. The embryos were then kept on 32°C/89.6°F as GAL4 has a maximal activity at higher temperatures (Duffy 2002) and analyzed for tGFP fluorescence every hour. (MZ 16FA fluorescence stereomicroscope with planachromatic 0.8 x objective; EGFP-LP filter set, 30 x magnification, 10 sec. exposure time)

Detection of reporter gene expression

Comparison of reporter gene expression was done via whole mount *in-situ* hybridization of 0-72 hour egg collections at 25°C/77°F. They had been heat shocked and fixed 11 hours later. Whole-mount *in-situ* hybridizations were performed with probes of approximately the same size and the same concentration (tGFP: 770 bp; 220 ng/µl; DsRed: 740bp, 230 ng/µl; eyfp: 790 bp, 220 ng/µl; lacZ: 750 bp, 230 ng/µl). Staining time was the same for all *in-situ* hybridizations. Whole-mount *in-situ* hybridizations were performed according to established protocols (Tautz and Pfeifle 1989).

Results

Constructs based on *Drosophila* constructs fail to work

Tribolium beetles transgenic for EGFP under the control of the artificial 3xP3 enhancer-promoter element exhibit strong EGFP fluorescence in larval pupal and adult eyes and parts of the nervous system (Berghammer et al. 1999). Moreover, several *Drosophila* constructs have been shown to work in other species (Imamura et al. 2003; Ramos et al. 2006). Therefore we expected that 3xP3 would efficiently drive GAL4 expression in the eyes.

Hence, our first approach to adapt the GAL4/UAS system to *Tribolium* was to directly transfer the constructs tested in *Drosophila* (Brand and Perrimon 1993; Viktorinova and Wimmer 2007) to *Tribolium*. We used constructs based on the transactivator versions GAL4Δ and GAL4V-P16 (Figure 1 b, c) which were driven by 3xP3. Transgenic beetles for each of these constructs were crossed to beetles carrying a responder with LacZ under the control of UAST (Figure 1 f) or UASp (Brand and Perrimon 1993; Rorth 1998). We were not able to detect LacZ expression in the eye anlagen and we did not find any evidence for a strongly enhanced LacZ activity (not shown). Unfortunately, some endogenous LacZ activity is present in the eyes that does not allow us to detect minor differences in LacZ activity upon misexpression. In parallel, we used *Tc-giant* under the control of UAST (Figure 1 g) as alternative responder. However, the analysis of heat shocked offspring for cuticle phenotypes did not reveal significant differences to the not heat shocked control (not shown).

GAL4Δ activates reporter gene expression faster than GAL4-VP16

As alternative approach we used *Tribolium* specific basal promoters in both the transactivator and the responder constructs. In the course of another project we have shown that a 150 bp fragment of the basal *hsp68* promoter shows sequence similarities to the *Drosophila* basal heat shock promoter and does not show activity itself but can be activated by endogenous and artificial heat shock elements (in preparation). In order to compare different versions of transactivators, we established transgenic *Tribolium* lines carrying either the GAL4Δ or the GAL4-VP16 version of the GAL4 activator (Ma and Ptashne 1987; Sadowski et al. 1988). Both transactivators are driven by the endogenous heat shock inducible Tc-hsp68 promoter element (in preparation) (Figure 1 d, e)

In the responder, the UAS sites from the Drosophila constructs were placed upstream of the Tc-hsp68 promoter and turboGFP (tGFP) including SV40 early mRNA polyadenylation signal was cloned downstream (Figure 1 k). In order to exclude position effects we analyzed two independent transgenic lines for each construct. Each of the four activator lines was crossed to two independent responder lines. Self crossed UAS responder lines were included as negative controls. A 24 hours egglay was collected and heat shocked (see materials and methods). Subsequently, the embryos were checked for fluorescence after 24 hours. The offspring of all but the negative controls showed strong tGFP expression (not shown). As the intensity of tGFP fluorescence was similar between the tested lines, we used the period of time from performing the heat shock to the onset of tGFP fluorescence as measure for transactivator efficiency. Earliest expression of tGFP in embryos could be observed using the UAS-Tc-bhsp-tGFP#2 line in combination with both GAL4Δ transactivator lines while the GAL4-VP16 lines took one hour more to show first tGFP fluorescence. The UAS-Tc-bhsp-tGFP#7 line tended to be activated later than the #2 line indicating some position effect. But again, GAL4Δ tended to perform better than GAL4-VP16. On average, when crossing the two different responder lines to the GAL4Δ lines tGFP fluorescence was visible 3.5 hours after heat shock whereas in crossings with the GAL4-VP16 lines 4.25 hours were necessary for first detection (Table 1).

GAL4/UAS is applicable during all stages of *Tribolium* development

To analyze whether the GAL4/UAS system is applicable in other stages and tissues in *Tribolium*, we performed heat shock in larvae, pupae and adult beetles. Directly and 24 hours after the heat shock treatment we checked for tGFP fluorescence. tGFP fluorescence was strongly increased ubiquitously in pupae and adults positive for both the GAL4Δ and UAS construct 24 h after heat shock (Figure 2d, h). Pupae and adults without heat shock (Figure 2b, f) or directly after heat shock (Figure 2c, g) and pupae as well as adults carrying either the transactivator (not shown) or the responder construct alone (Figure 2a, e) showed no increased fluorescence compared to wild type pupae or adults. The same was true for larvae (data not shown). This experiment demonstrates that the GAL4Δ system appears to be active throughout all stages of development. We screened several tissues of adults for reporter activity and find that the system also works in a variety of tissues. The wings (Figure 2 h), male (Figure 2 m) and female reproductive organs (Figure 2 q) as well as the gut (Figure 2 u) show strong fluorescence of tGFP 24 hours after heat shock. Control beetles of the same genotype do not exhibit fluorescence in any of these tissues without heat shock (Figure 2 k, o, s).

Endogenous promoters are required for efficient function of both transactivator and responder

Previously, several *Drosophila* promoters have been used in the transactivator and responder construct to establish the GAL4/UAS binary expression system in *Tribolium* without any success. Our results suggest that the use of the endogenous basal promoter is critical for the function. Therefore, we wanted to compare the relative efficiencies of different basal promoters by using our tested functional GAL4Δ transactivator and UAS responder lines. We crossed the functional driver line Tc-hsp-GAL4Δ#1 to different responder lines based on non-*Tribolium* basal promoters (Figure 1 f, h), induced GAL4Δ expression by a heat shock and detected the transcript of the reporter gene. For a comparable staining, all probes were approximately the same size, adjusted to approximately the same concentration and the stainings were develped the same time. Two independent insertion sites for each responder construct were analyzed to control for integration site effects, respectively. As positive control the lines UAS-Tc-bhsp-tGFP#2 and #7 were crossed to the same driver line. As expected, these embryos of the positive control show strong expression of the reporter gene tGFP (Figure 3a, b).

First, we tested a responder construct containing the *Drosophila* specific basal *hsp70* promoter (Dm-hsp70TATA) which has been used in *Drosophila* to drive LacZ (UAS-Dm-hsp-LacZ #MIII and #MII, Figure 1 f) We did not detect expression of tGFP in the offspring (Figure 3c, d).

Next, we tested the SCP1 basal promoter which is an artificial basal promoter that contains four core promoter motifs - the TATA box from CytoMegalie-Virus (CMV) IE1, the initiator (Inr) is based on Adnovirus Major Late (AdML) and *Drosophila* G retrotransposon, motif ten element (MTE) from *Drosophila Tollo* and the downstream promoter element (DPE) from *Drosophila* G core promoter- in a single promoter. It directs high amounts of transcription by RNA polymerase II in nuclear extracts from *Drosophila* and HeLa cells and is more efficient than the CMV or AdML core promoters (Juven-Gershon et al. 2006). This construct contains additional GAGA elements to prevent position effects (O'Donnell and Wensink 1994; Tsukiyama et al. 1994). However, when the lines UAS-SCP1-DsRed #F2 and #M3 (Figure 1 h) were crossed to our driver line and were heat shocked no expression of the reporter gene *DsRed* could be observed in offspring embryos (Figure 3e, f). These results indicate that neither a basal promoter from *Drosophila* nor an artificial promoter comprising four core promoter motifs is efficient in driving expression in *Tribolium*.

Assuming that endogenous basal promoters are required for efficient transcription, we tested another *Tribolium* basal promoter. The *hairy* upstream region has been analyzed previously and from these data, the putative *Tc-hairy* basal promoter was deduced (Eckert et al. 2004). We added this putative basal *Tc'hairy* promoter (*Tc-h_p*) in the UAS-Dm-hsp-LacZ construct and exchanged LacZ by EYFP just downstream of the Dm-hsp70TATA generating a construct with both *Drosophila* and *Tribolium* basal promoters (Figure 1 i). When we crossed the lines UAS-Dm-hsp-Tc-bh-EYFP #3.3 and #6 against the driver line strong expression of the reporter gene *eyfp* could be detected predominantly within the central nervous system in old embryos close to dorsal closure (Figure 3g, h). At earlier stages expression was not efficient. As this surprising expression pattern was identical in two independently generated insertions of the same construct and as *Tc-hairy* is not expressed in the central nervous system in embryos of this developmental stage (not shown), this unexpected restriction to the nervous system is probably neither due to the integration site nor to the *Tc-hairy* basal promoter but is a property of the construct. The negative control (UAS-Dm-hsp-Tc-bh-EYFP alone) does not exhibit detectable expression of *eyfp* in the absence of GAL4 transactivator activity (not shown). Apparently, the construct responds to GAL4 predominantly in a subset of cells.

In the same way we also tested the efficiency of *Drosophila* promoters in transactivator lines which we mated with the responder UAS-Tc-bhsp-tGFP #2. Dm-hs-GAL4 #15 and #16 (Figure 1 a) are based on the upstream region of the *Drosophila* heat shock gene, *hsp70* (Lis et al. 1983). These heat shock constructs are frequently used in *Drosophila* (e.g. (Certel and Johnson 1996; Handler and Harrell 1999)). The transactivator line Dm-hs-GAL4 #16 induced weak expression of the reporter gene tGFP in single cells of very old embryos (Figure 3 k) whereas in the mating with transactivator line Dm-hs-GAL4 #15 no tGFP expression was detected (Figure 3 i). This shows that apparently, there is some weak activity of the *Drosophila* heat shock promoter in *Tribolium* which appears to depend highly on the integration site. This is confirmed by the testing of several different *Drosophila* heat shock constructs in *Tribolium* which, depending on the integration site, are inducible to varying degrees and in different tissues (Viktorinova and Wimmer, unpublished).

Temperature dependence of GAL4/UAS system

In *Drosophila* it has been shown that GAL4 activity is enhanced in flies raised at 29°C/84.2°F compared to lower temperatures (Brand et al. 1994). Therefore we wanted to test if this temperature dependence exists in *Tribolium* as well. 0 to 24 hours old *Tribolium* embryos derived from heterozygous parents carrying the transactivator Tc-hsp-GAL4Δ#1 and the responder UAS-Tc-bhsp-tGFP#7 were heat shocked, dechorionated, aligned on a microscope slide and kept at different temperatures. Pictures were taken directly after the heat shock and in one hour intervals to check for the onset of tGFP fluorescence. Indeed we find that higher temperatures lead to an earlier onset of fluorescence. When Embryos were kept at 26°C/78.8°F five hours were necessary until first tGFP fluorescence was visible, at 28°C/82.4°F four to five hours, at 30°C/86°F three to four hours and at 32°C/89.6°F three hours, respectively (Figure 4). As a negative control, embryos carrying only the responder line were heat shocked and kept at 30°C/86°F. Even 24 hours after heat shock no tGFP fluorescence was observed (Figure 4Eg).

The earlier fluorescence of tGFP at higher temperatures is probably mainly due to the faster formation of the GAL4Δ as there is no major difference between the efficiency of chromophore formation at 28°C and 37°C for several GFP variants. This efficiency even decreases for some variants at higher temperatures (Patterson et al. 1997). Taken these data together, the transactivation by GAL4Δ is faster at 32°C/89.6°F than at lower temperatures. We did not test higher temperatures because *Tribolium* should not be kept at temperatures higher than 32°C/89.6°F.

GAL4/UAS system is applicable in *Tribolium* embryogenesis

We furthermore analyzed if the GAL4/UAS response is fast enough to analyze gene function during embryogenesis. This is only possible if there is only a short delay between activation of the driver and expression of the responder compared to embryonic development. We tested how much time it takes from the heat shock activation of the driver to expression of the reporter gene of the responder using the previously determined best combination (Tc-hsp-GAL4Δ#1 and UAS-Tc-bhsp-tGFP #7). A one hour egglay was collected and kept until the end of elongation (18-19h at 32°C). The embryos were heat shocked and fixed for *in-situ* hybridization at different time points after heat shock. It was unclear how temperature would affect the system as embryonic development is about twice as fast at 32°C compared to 25°C but maturation of GAL4 is faster at higher temperatures. Hence, treated embryos were kept at 32°C or 25°C after heat shock. Embryos kept at 25°C were fixed directly and two, four and six hours after heat shock treatment (Figure 5 a-d). Embryos kept at 32°C were fixed directly after heat shock and after two, three and four hours because faster onset was expected at higher temperatures (see above) (Figure 5 n-q). These embryos were then checked for tGFP expression via *in-situ* hybridization. Earliest tGFP expression could be observed four hours after heat shock at 25°C, whereas embryos kept at 32°C showed expression already after two hours (compare Figure 5 c to o). Strong expression was detectable already after three hours at 32°C while six hours were necessary for a similar expression level at 25°C (compare Figure 5 d to p). At the stages tested, the morphology of the head and the elongation of appendages provide a good measure for the age of the embryo. We compared both overall morphology, head morphology and the length of the trunk appendages at the time of heat shock (Figure 5 e, i and r, v) and when the reporter was fully expressed (Figure 5 h, m and t, x, respectively) for at least 5 embryos per time point. We do not find a major difference indicating that the response is very fast relative to the developmental time.

Discussion

We have adopted the GAL4/UAS system to *Tribolium* and find that GAL4Δ (Ma and Ptashne 1987) is slightly superior to GAL4-VP16 (Sadowski et al. 1988). This came not unexpected as GAL4Δ is a much smaller protein that consists only of the DNA binding and activation domain of GAL4. As toxic effects of GAL4-VP16 have been observed for *Drosophila* (Driever et al. 1989; Viktorinova and Wimmer 2007) we suggest to use GAL4Δ in *Tribolium* in the future. For highest activity the experiments should be carried out at 32°C/89.6°F.

The GAL4Δ/UAS system is active at all stages of *Tribolium* development and activates reporter gene expression in a variety of different tissues indicating broad applicability. Interestingly, the activation of a reporter is very fast relative to embryonic development which will allow using the system to investigate this process. In contrast, the extremely fast mode of *Drosophila* early development has in many cases hampered its use in the study of these processes. Because of the higher temperature *Tribolium* is adapted to, the system is likely to be even more active than in *Drosophila*.

One important issue to consider for the design of future transgenic tools comes from our comparison of functionality of different basal promoters. Neither the *Drosophila* basal heat shock promoter nor an artificial "super core promoter" consisting of four core promoter motifs were effective in driving expression in *Tribolium* while endogenous promoters were. This finding was somewhat unexpected as *Drosophila* heat shock constructs have been shown to work in the silkworm *Bombyx mori* (Imamura et al. 2003) and butterfly *Bicyclus anynana* (Ramos et al. 2006). Moreover, the artificial 3xP3 enhancer is driving a reporter in the *Tribolium* eyes although it is based on a *Drosophila* basal promoter (Berghammer et al. 1999). Also *Drosophila* heat shock constructs have been shown to work in *Tribolium* but their efficiency appears to be highly dependent on the integration site (Wimmer, personal communication). Taken together, although exogenous promoters may work to some extent in *Tribolium* under certain circumstances, for full functionality, the use of species-specific promoters in *Tribolium* is essential.

In order to provide full benefit of the system, it is essential to generate a collection of driver lines that can then be used to misexpress genes in certain tissues. One way is a random insertion screen with a GAL4Δ driver. Therefore we have constructed a *piggy*Bac transposon containing the GAL4Δ sequence under the control of the basal *Tc-hsp68* heat shock promoter. This construct does not express GAL4Δ unless it integrates close to an enhancer. With this mutator we are currently initiating an enhancer trap screen similar to the one described in chapter 3.2.1 (Trauner et al., 2009; submitted) in order to establish new transactivator lines.

The binary expression system GAL4/UAS is an utmost useful tool to unravel gene function in different scientific contexts and in different species. With its establishment in *Tribolium*, we open additional experimental possibilities to study gene function. This will further foster *Tribolium* as a model organism where primary questions can be studied.

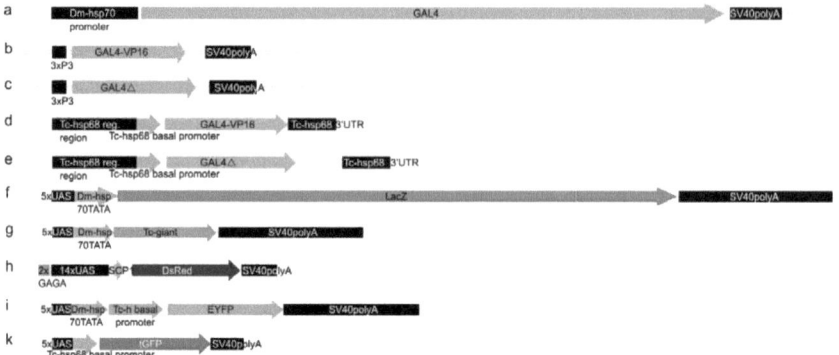

Figure 1: Schematic representation of responder and transactivator constructs. (a) Dm-hs-GAL4; (b) 3xP3-GAL4-VP16; (c) 3xP3-GAL4Δ; (d) Tc-hsp-GAL4-VP16; (e) Tc-hsp-GAL4Δ; (f) UAS-Dm-hsp-LacZ; (g) UAS-Dm-hsp-Tc-giant; (h) UAS-SCP1-DsRed; this construct contains GAGA sites to prevent position effects. (i) UAS-Dm-hsp-Tc-bh-EYFP; (k) UAS-Tc-bhsp-tGFP

Table 1: Comparison of GAL4Δ and GAL4-VP16

	UAS-Tc-bhsp-tGFP	
	#2	#7
Tc-hsp-GAL4Δ #1	3 h	4 h
Tc-hsp-GAL4Δ #2	3 h	4 h
Tc-hsp-GAL4-VP16 #2	4 h	4 h
Tc-hsp-GAL4-VP16 #3	4 h	6 h

GAL4Δ is slightly faster than GAL4-VP16. Shown is the time in hours until first fluorescence of tGFP was visible in the respective crossings. The transactivator lines Tc-hsp-GAL4Δ#1 and #2 and Tc-hsp-GAL4-VP16#2 and #3 were crossed against the responder lines UAS-Tc-bhsp-tGFP#2 and #7. Transactivator lines containing GAL4Δ lead to an earlier responder gene expression (3.5 hours on average) than lines containing GAL4-VP16 (4.25 hours on average).

The binary expression system GAL4/UAS-Manuscript

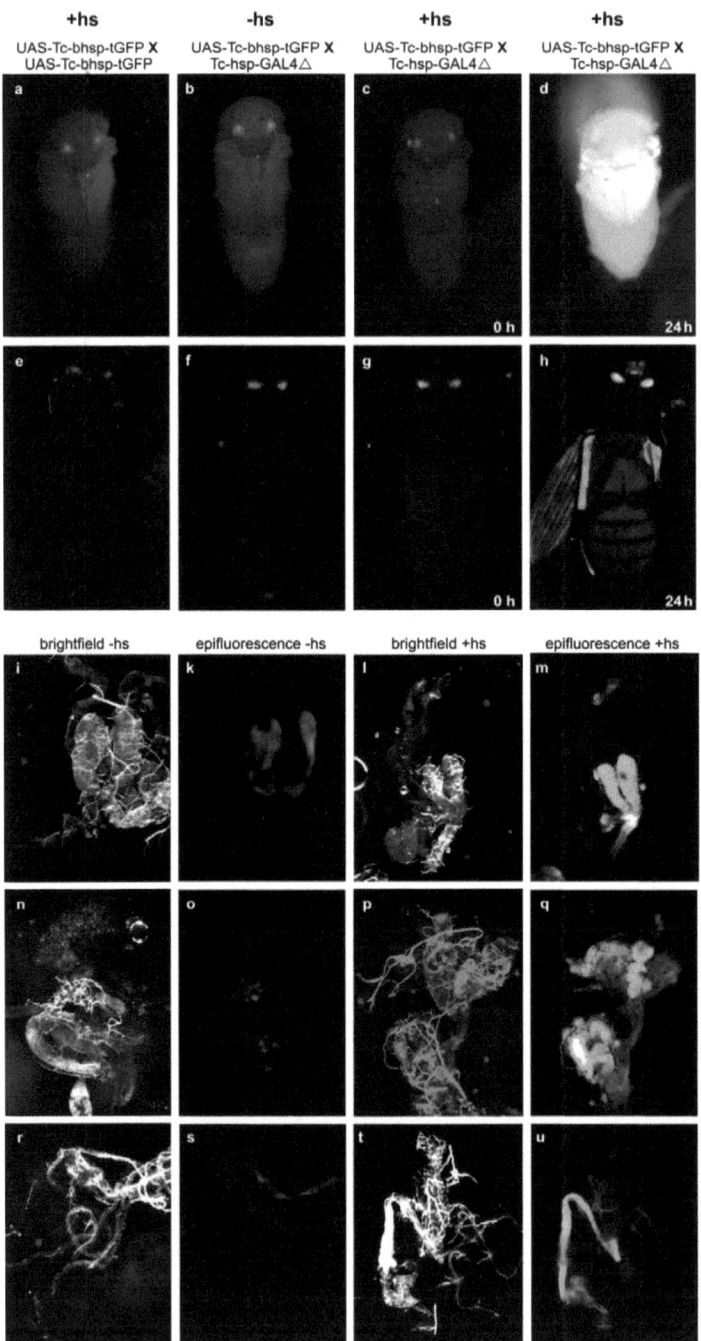

Figure 2: Application of GAL4Δ/UAS system at different stages and tissues of *Tribolium*. Only pupae and adults positive for both, the GAL4Δ driver and UAS responder construct show strong tGFP fluorescence after heat shock. Untreated pupae and adults or *Tribolium* positive for only one of the constructs show no fluorescence. (**a, e**) Negative control: No fluorescence is visible 24 h after heat shock in animals that carry only the responder line UAS-Tc-bhsp-tGFP#7 (marked by red eyes only) (**b, f**) Animals carrying driver (Tc-hsp-GAL4Δ#1) and responder (UAS-Tc-bhsp-tGFP#7) construct (marked by red and blue eyes), handled like the others, but without heat shock show no fluorescence of tGFP. (**c, g**) Animals carrying driver and responder construct (red and blue eyes) directly after heat shock exhibit no fluorescence of tGFP. (**d, h**) Animals 24 h after heat shock exhibit strong tGFP fluorescence. (**h**) The wings also exhibit tGFP fluorescence. (i, n, r and l, q, t) show brightfield pictures of (k, o, s and m, q, u). Tissues of beetles carrying driver and responder construct but without heat shock exhibit no fluorescence of tGFP in (**k**) male reproductive organs, (**o**) female reproductive organs, (**s**) gut, whereas 24 hours after heat shock strong fluorescence is visible in the respective tissues (**m, q, u**).

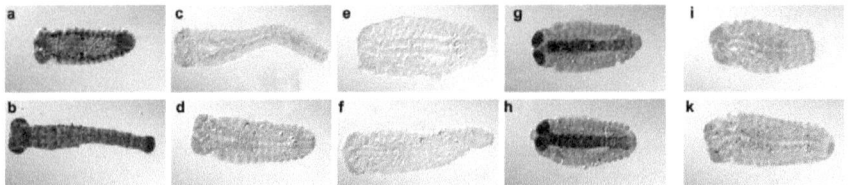

Figure 3: Endogenous versus exogenous promoters in transactivator and responder constructs. Only when *Tribolium* specific basal promoters are used in the responder and transactivator constructs, activation of the reporter gene is observed via *in-situ* hybridization. Exogenous promoters were not capable of driving reporter gene expression. (a-h)Transactivator line Tc-hsp-GAL4Δ#1 was crossed to different responder lines and *in-situ* hybridization was performed with the respective antisense RNA probe. Transactivator line crossed to: (**a**) UAS-Tc-bhsp-tGFP#2 and *in-situ* with tGFP antisense probe as positive control; ubiquitous expression of tGFP; (**b**) UAS-Tc-bhsp-tGFP#7 and *in-situ* with tGFP antisense probe as positive control; ubiquitous expression of tGFP; (**c**) UAS-Dm-hsp-LacZ#MIII and *in-situ* with LacZ antisense probe; no expression of LacZ; (**d**) UAS-Dm-hsp-LacZ#MII and *in-situ* with LacZ antisense probe; no expression of LacZ; (**e**) UAS-SCP1-DsRed #F2 and *in-situ* with DsRed antisense probe; no expression of DsRed; (**f**) UAS-SCP1-DsRed #M3 and *in-situ* with DsRed antisense probe; no expression of DsRed; (**g**) UAS-Dm-hsp-Tc-bh-EYFP#3.3 and *in-situ* with EYFP antisense probe; strong expression of EYFP in the central nervous system; (**h**) UAS-Dm-hsp-Tc-bh-EYFP#6 and *in-situ* with EYFP antisense probe; strong expression of EYFP in the central nervous system. (**i, k**) Responder line UAS-Tc-bhsp-tGFP#2 crossed to: (**i**) Dm-hs-GAL4#15 and *in-situ* with tGFP antisense probe; no expression of tGFP (**k**) Dm-hsp-GAL4#16 and *in-situ* with tGFP antisense probe; weak expression of tGFP in single cells of the embryo.

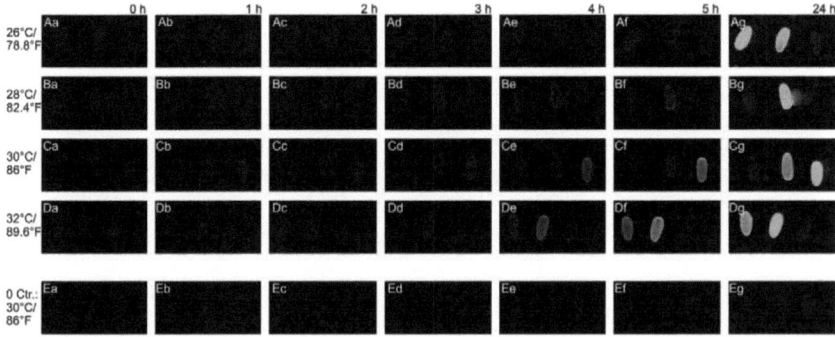

Figure 4: Temperature dependence of GAL4/UAS system. The GAL4/UAS system is temperature dependent and shows higher activity at higher temperatures. The transactivator line Tc-hsp-GAL4Δ#1 was crossed to the responder line UAS-Tc-bhsp-tGFP#7. Embryos of this crossing were checked for tGFP fluorescence directly after heat shock and in one hour intervals. Rows correspond to different temperatures (first row 26°C, second 28°C, third 30°C, fourth 32°C). As both lines are heterozygous only about 25% of embryos were expected to exhibit tGFP fluorescence. In the bottom row embryos of the responder line alone at 30°C are shown as negative control. Each row shows the same embryos at a given time after heat shock. At 26°C first fluorescence is seen 5 hours after heat shock (**Af**); 28°C: 4-5 hours (**Be; Bf**); 30°C: 3-4 hours (**Cd; Ce**); 32°C: 3 hours (**Dd**). No tGFP fluorescence was visible in embryos of the responder line alone even 24 hours after heat shock (**Eg**). Note that only part of the offspring carries both constructs because both parents were heterozygous.

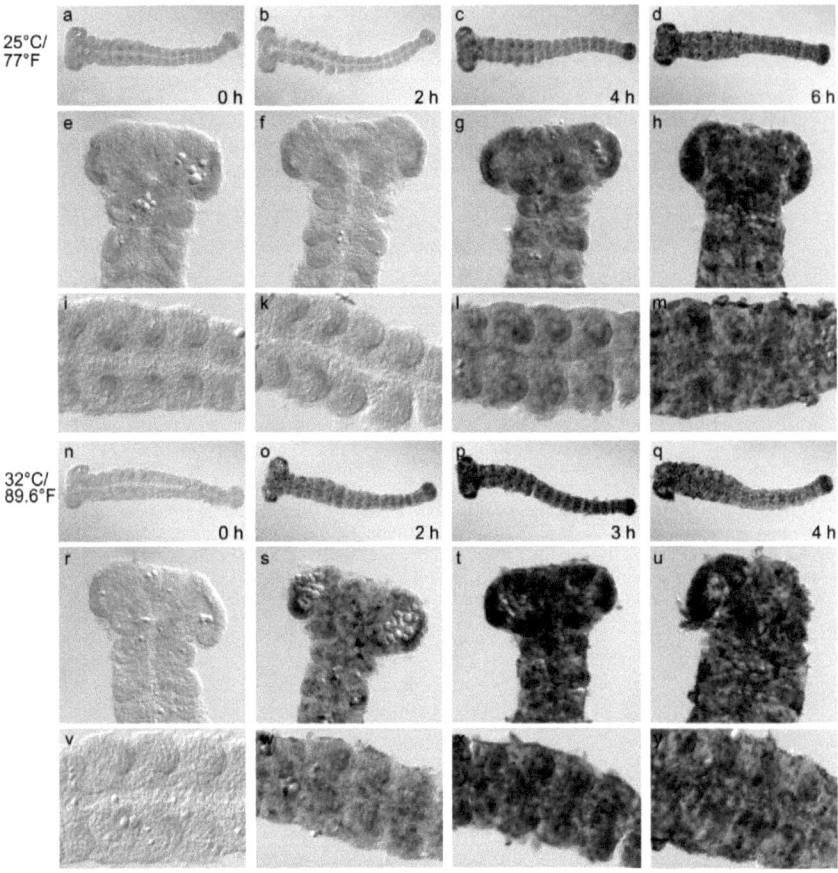

Figure 5: Onset of GAL4 driven expression is fast relative to embryonic development. Offspring of line Tc-hsp-GAL4Δ #1 crossed to line UAS-Tc-bhsp-tGFP#7. 18 hours to 19 hours old embryos were collected and heat shocked. The embryos were kept at 25°C (a-d) and 32°C (n-q) respectively, after heat shock treatment, fixed at different time points and checked for tGFP expression via in-situ: (a, b) There is no expression of reporter gene tGFP two hours after heat shock. (c) After 4 h weak expression is detected (d) After 6 h strong expression of the reporter gene is seen; (n) No expression directly after heat shock. (o) Onset of expression is seen after 2 h. (p, q) From 3 h after heat shock on, strong expression is observed. (e-h) and (r-u) show higher magnification of the heads, (i-m) and (v-y) of the thorax of the respective embryos. The developmental stage of the embryos directly after heat shock and at the time point of first tGFP expression was determined by the outgrowth of head and thoracic appendages. There is hardly any difference observable (compare e, i to g, l and r, v to s, w).

References

Berghammer, A. J., M. Klingler, et al. (1999). "A universal marker for transgenic insects." Nature **402**(6760): 370-1.

Brand, A. H., A. S. Manoukian, et al. (1994). "Ectopic expression in Drosophila." Methods Cell Biol **44**: 635-54.

Brand, A. H. and N. Perrimon (1993). "Targeted gene expression as a means of altering cell fates and generating dominant phenotypes." Development **118**(2): 401-15.

Brown, S. J., J. P. Mahaffey, et al. (1999). "Using RNAi to investigate orthologous homeotic gene function during development of distantly related insects." Evol Dev **1**(1): 11-5.

Bucher, G., J. Scholten, et al. (2002). "Parental RNAi in Tribolium (Coleoptera)." Curr Biol **12**(3): R85-6.

Cerny, A. C., D. Grossmann, et al. (2008). "The Tribolium ortholog of knirps and knirps-related is crucial for head segmentation but plays a minor role during abdominal patterning." Dev Biol **321**(1): 284-94.

Cormack, B. P., R. H. Valdivia, et al. (1996). "FACS-optimized mutants of the green fluorescent protein (GFP)." Gene **173**(1 Spec No): 33-8.

Cubitt, A. B., L. A. Woollenweber, et al. (1999). "Understanding structure-function relationships in the Aequorea victoria green fluorescent protein." Methods Cell Biol **58**: 19-30.

Driever, W., J. Ma, et al. (1989). "Rescue of bicoid mutant Drosophila embryos by bicoid fusion proteins containing heterologous activating sequences." Nature **342**(6246): 149-54.

Duffy, J. B. (2002). "GAL4 system in Drosophila: a fly geneticist's Swiss army knife." Genesis **34**(1-2): 1-15.

Eckert, C., M. Aranda, et al. (2004). "Separable stripe enhancer elements for the pair-rule gene hairy in the beetle Tribolium." EMBO Rep **5**(6): 638-42.

Fischer, J. A., E. Giniger, et al. (1988). "GAL4 activates transcription in Drosophila." Nature **332**(6167): 853-6.

Giniger, E., S. M. Varnum, et al. (1985). "Specific DNA binding of GAL4, a positive regulatory protein of yeast." Cell **40**(4): 767-74.

Guyer, D., A. Tuttle, et al. (1998). "Activation of latent transgenes in Arabidopsis using a hybrid transcription factor." Genetics **149**(2): 633-9.

Handler, A. M. and R. A. Harrell, 2nd (1999). "Germline transformation of Drosophila melanogaster with the piggyBac transposon vector." Insect Mol Biol **8**(4): 449-57.

Hartley, K. O., S. L. Nutt, et al. (2002). "Targeted gene expression in transgenic Xenopus using the binary Gal4-UAS system." Proc Natl Acad Sci U S A **99**(3): 1377-82.

Horn, C., B. G. Schmid, et al. (2002). "Fluorescent transformation markers for insect transgenesis." Insect Biochem Mol Biol **32**(10): 1221-35.

Horn, C. and E. A. Wimmer (2000). "A versatile vector set for animal transgenesis." Dev Genes Evol **210**(12): 630-7.

Hunt, T., J. Bergsten, et al. (2007). "A comprehensive phylogeny of beetles reveals the evolutionary origins of a superradiation." Science **318**(5858): 1913-6.

Imamura, M., J. Nakai, et al. (2003). "Targeted gene expression using the GAL4/UAS system in the silkworm Bombyx mori." Genetics **165**(3): 1329-40.

Juven-Gershon, T., S. Cheng, et al. (2006). "Rational design of a super core promoter that enhances gene expression." Nat Methods **3**(11): 917-22.

Konopova, B. and M. Jindra (2007). "Juvenile hormone resistance gene Methoprene-tolerant controls entry into metamorphosis in the beetle Tribolium castaneum." Proc Natl Acad Sci U S A **104**(25): 10488-93.

Konopova, B. and M. Jindra (2008). "Broad-Complex acts downstream of Met in juvenile hormone signaling to coordinate primitive holometabolan metamorphosis." Development **135**(3): 559-68.

Laughon, A., R. Driscoll, et al. (1984). "Identification of two proteins encoded by the Saccharomyces cerevisiae GAL4 gene." Mol Cell Biol **4**(2): 268-75.

Laughon, A. and R. F. Gesteland (1984). "Primary structure of the Saccharomyces cerevisiae GAL4 gene." Mol Cell Biol **4**(2): 260-7.

Lorenzen, M. D., A. J. Berghammer, et al. (2003). "piggyBac-mediated germline transformation in the beetle Tribolium castaneum." Insect Mol Biol **12**(5): 433-40.

Lorenzen, M. D., T. Kimzey, et al. (2007). "piggyBac-based insertional mutagenesis in Tribolium castaneum using donor/helper hybrids." Insect Mol Biol **16**(3): 265-75.

Ma, J. and M. Ptashne (1987). "The carboxy-terminal 30 amino acids of GAL4 are recognized by GAL80." Cell **50**(1): 137-42.

Ma, J. and M. Ptashne (1987). "Deletion analysis of GAL4 defines two transcriptional activating segments." Cell **48**(5): 847-53.

McGuire, S. E., G. Roman, et al. (2004). "Gene expression systems in Drosophila: a synthesis of time and space." Trends Genet **20**(8): 384-91.

McKnight, J. L., T. M. Kristie, et al. (1987). "Binding of the virion protein mediating alpha gene induction in herpes simplex virus 1-infected cells to its cis site requires cellular proteins." Proc Natl Acad Sci U S A **84**(20): 7061-5.

Miller, S. C., S. J. Brown, et al. (2008). "Larval RNAi in Drosophila?" Dev Genes Evol **218**(9): 505-10.

O'Hare, P. and C. R. Goding (1988). "Herpes simplex virus regulatory elements and the immunoglobulin octamer domain bind a common factor and are both targets for virion transactivation." Cell **52**(3): 435-45.

O'Hare, P., C. R. Goding, et al. (1988). "Direct combinatorial interaction between a herpes simplex virus regulatory protein and a cellular octamer-binding factor mediates specific induction of virus immediate-early gene expression." Embo J **7**(13): 4231-8.

Ornitz, D. M., R. W. Moreadith, et al. (1991). "Binary system for regulating transgene expression in mice: targeting int-2 gene expression with yeast GAL4/UAS control elements." Proc Natl Acad Sci U S A **88**(3): 698-702.

Patterson, G., R. N. Day, et al. (2001). "Fluorescent protein spectra." J Cell Sci **114**(Pt 5): 837-8.

Patterson, G. H., S. M. Knobel, et al. (1997). "Use of the green fluorescent protein and its mutants in quantitative fluorescence microscopy." Biophys J **73**(5): 2782-90.

Preston, C. M., M. C. Frame, et al. (1988). "A complex formed between cell components and an HSV structural polypeptide binds to a viral immediate early gene regulatory DNA sequence." Cell **52**(3): 425-34.

Richards, S., R. A. Gibbs, et al. (2008). "The genome of the model beetle and pest Tribolium castaneum." Nature **452**(7190): 949-55.

Rorth, P. (1998). "Gal4 in the Drosophila female germline." Mech Dev **78**(1-2): 113-8.

Sadowski, I., J. Ma, et al. (1988). "GAL4-VP16 is an unusually potent transcriptional activator." Nature **335**(6190): 563-4.

Scheer, N. and J. A. Campos-Ortega (1999). "Use of the Gal4-UAS technique for targeted gene expression in the zebrafish." Mech Dev **80**(2): 153-8.

Stebbins, M. J. and J. C. Yin (2001). "Adaptable doxycycline-regulated gene expression systems for Drosophila." Gene **270**(1-2): 103-11.

Suzuki, Y., D. C. Squires, et al. (2009). "Larval leg integrity is maintained by Distal-less and is required for proper timing of metamorphosis in the flour beetle, Tribolium castaneum." Dev Biol **326**(1): 60-7.

Tautz, D. and C. Pfeifle (1989). "A non-radioactive in situ hybridization method for the localization of specific RNAs in Drosophila embryos reveals translational control of the segmentation gene hunchback." Chromosoma **98**(2): 81-5.

Tomoyasu, Y. and R. E. Denell (2004). "Larval RNAi in Tribolium (Coleoptera) for analyzing adult development." Dev Genes Evol **214**(11): 575-8.

Tomoyasu, Y., S. C. Miller, et al. (2008). "Exploring systemic RNA interference in insects: a genome-wide survey for RNAi genes in Tribolium." Genome Biol **9**(1): R10.

Triezenberg, S. J., R. C. Kingsbury, et al. (1988). "Functional dissection of VP16, the trans-activator of herpes simplex virus immediate early gene expression." Genes Dev **2**(6): 718-29.

Triezenberg, S. J., K. L. LaMarco, et al. (1988). "Evidence of DNA: protein interactions that mediate HSV-1 immediate early gene activation by VP16." Genes Dev **2**(6): 730-42.

Viktorinova, I. and E. A. Wimmer (2007). "Comparative analysis of binary expression systems for directed gene expression in transgenic insects." Insect Biochem Mol Biol **37**(3): 246-54.

Yang, T. T., L. Cheng, et al. (1996). "Optimized codon usage and chromophore mutations provide enhanced sensitivity with the green fluorescent protein." Nucleic Acids Res **24**(22): 4592-3.

4 General Discussion

With this thesis I have contributed to the establishment of *Tribolium castaneum* as a model system for insect head development. The analysis of the *Tribolium* orthologs of the head gap-like genes of *Drosophila* revealed that the early blastodermal function and expression is highly variable between different species (Schinko et al. 2008). Also Bicoid is only found in higher Dipterans and it is suggested that its function is substituted by Orthodenticle and Hunchback (Lynch and Desplan 2003; Schroder 2003). These findings suggest a fast change in early patterning in insects (Brown et al. 2001; Schroder 2003). Due to this and the fact that the mode of larval head development is derived, *Drosophila* is less suitable for analyzing head patterning and findings in this species are not necessarily typical for insects. As *Tribolium* larval head development is typical for insects and robust gene knock down is established, many candidate genes from *Drosophila* or other species can be analyzed to get deeper insights into the process of head patterning.

No comprehensive list of genes required for head development in *Drosophila* is available. Therefore a hypothesis independent approach was necessary to identify novel genes involved in this process. The GEKU screen provided many new starting-points for further analysis of head development. Up to now only seven out of the 39 interesting lethal lines have been analyzed in some more detail. Of those, four appear to be especially interesting (G07411, G09104, KS0294 and KT1269). It has been shown that for some of the lethal lines it was straight forward to identify the probably affected gene and confirm this via RNAi. Problems occurred if the probably affected genes were necessary for viability or fertility or were essential during embryonic development. In these cases parental RNAi did not lead to any results. By performing embryonic RNAi the adult lethality and sterility can be circumvented and the embryonic defects can be analyzed and compared to the mutant phenotype. If the embryonic RNAi phenocopies the mutant phenotype the affected gene is confirmed and offspring of the mutant lines can be used for further analysis via *in-situ* hybridization as embryonic RNAi for *in-situ* hybridization is utmost laborious and time consuming. Thus, the function of these genes can be analyzed less laboriously with the mutant lines in hand.

General Discussion

The work described in this thesis concerning the GEKU insertional mutagenesis screen is one of the initial steps to get deeper insights into the complex procedure of arthropod head development. It shows that novel genes are identified by forward genetics and, more specifically, several genes have been identified for further analysis.

In order to understand the function of newly identified or already known genes not only knock down but also misexpression is desirable. Hence I developed the binary misexpression system GAL4/UAS for *Tribolium*. Several trials in the previous years have been not successful. The main difference to the previous approaches was that I used *Tribolium* specific promoters in contrast to the previously used *Drosophila* specific ones. This leads to a functional system that will be a powerful tool once a collection of drivers is available. To this end, a GAL4Δ construct with the *Tc-hsp68* basal promoter should be used for an enhancer trap screen. More generally, this indicates that for transgenic approaches in *Tribolium* the use of endogenous promoters is necessary although *Drosophila* specific constructs have been used successfully in other species.

Due to all the previously established techniques in combination with the genome sequenced and the work done in this thesis, *Tribolium* has evolved into the prime model organism for insect head development.

5 References

Angelini, D. R., M. Kikuchi, et al. (2009). "Genetic patterning in the adult capitate antenna of the beetle Tribolium castaneum." Dev Biol **327**(1): 240-51.
Arakane, Y., S. Muthukrishnan, et al. (2005). "The Tribolium chitin synthase genes TcCHS1 and TcCHS2 are specialized for synthesis of epidermal cuticle and midgut peritrophic matrix." Insect Mol Biol **14**(5): 453-63.
Bello, B., D. Resendez-Perez, et al. (1998). "Spatial and temporal targeting of gene expression in Drosophila by means of a tetracycline-dependent transactivator system." Development **125**(12): 2193-202.
Berghammer, A., G. Bucher, et al. (1999). "A system to efficiently maintain embryonic lethal mutations in the flour beetle Tribolium castaneum." Dev Genes Evol **209**(6): 382-9.
Berghammer, A. J., M. Klingler, et al. (1999). "A universal marker for transgenic insects." Nature **402**(6760): 370-1.
Brand, A. H., A. S. Manoukian, et al. (1994). "Ectopic expression in Drosophila." Methods Cell Biol **44**: 635-54.
Brand, A. H. and N. Perrimon (1993). "Targeted gene expression as a means of altering cell fates and generating dominant phenotypes." Development **118**(2): 401-15.
Briscoe, J., A. Pierani, et al. (2000). "A homeodomain protein code specifies progenitor cell identity and neuronal fate in the ventral neural tube." Cell **101**(4): 435-45.
Brown, S., J. Fellers, et al. (2001). "A strategy for mapping bicoid on the phylogenetic tree." Curr Biol **11**(2): R43-4.
Brown, S. J., J. P. Mahaffey, et al. (1999). "Using RNAi to investigate orthologous homeotic gene function during development of distantly related insects." Evol Dev **1**(1): 11-5.
Bucher, G., J. Scholten, et al. (2002). "Parental RNAi in Tribolium (Coleoptera)." Curr Biol **12**(3): R85-6.
Bucher, G. and E. A. Wimmer (2005). "Beetle a-head." B.I.F. Futura **20**: 164–169.
Cerny, A. C., D. Grossmann, et al. (2008). "The Tribolium ortholog of knirps and knirps-related is crucial for head segmentation but plays a minor role during abdominal patterning." Dev Biol **321**(1): 284-94.
Certel, S. J. and W. A. Johnson (1996). "Disruption of mesectodermal lineages by temporal misexpression of the Drosophila POU-domain transcription factor, drifter." Dev Genet **18**(4): 279-88.
Chittaranjan, S., M. McConechy, et al. (2009). "Steroid hormone control of cell death and cell survival: molecular insights using RNAi." PLoS Genet **5**(2): e1000379.
Cohen, S. and G. Jurgens (1991). "Drosophila headlines." Trends Genet **7**(8): 267-72.
Cohen, S. M. and G. Jurgens (1990). "Mediation of Drosophila head development by gap-like segmentation genes." Nature **346**(6283): 482-5.
Cormack, B. P., R. H. Valdivia, et al. (1996). "FACS-optimized mutants of the green fluorescent protein (GFP)." Gene **173**(1 Spec No): 33-8.
Cubitt, A. B., L. A. Woollenweber, et al. (1999). "Understanding structure-function relationships in the Aequorea victoria green fluorescent protein." Methods Cell Biol **58**: 19-30.
Dalton, D., R. Chadwick, et al. (1989). "Expression and embryonic function of empty spiracles: a Drosophila homeo box gene with two patterning functions on the anterior-posterior axis of the embryo." Genes Dev **3**(12A): 1940-56.

References

Deak, P., M. Donaldson, et al. (2003). "Mutations in makos, a Drosophila gene encoding the Cdc27 subunit of the anaphase promoting complex, enhance centrosomal defects in polo and are suppressed by mutations in twins/aar, which encodes a regulatory subunit of PP2A." J Cell Sci **116**(Pt 20): 4147-58.

Driever, W., J. Ma, et al. (1989). "Rescue of bicoid mutant Drosophila embryos by bicoid fusion proteins containing heterologous activating sequences." Nature **342**(6246): 149-54.

Duffy, J. B. (2002). "GAL4 system in Drosophila: a fly geneticist's Swiss army knife." Genesis **34**(1-2): 1-15.

Eckert, C., M. Aranda, et al. (2004). "Separable stripe enhancer elements for the pair-rule gene hairy in the beetle Tribolium." EMBO Rep **5**(6): 638-42.

Economou, A. D. and M. J. Telford (2009). "Comparative gene expression in the heads of Drosophila melanogaster and Tribolium castaneum and the segmental affinity of the Drosophila hypopharyngeal lobes." Evol Dev **11**(1): 88-96.

Erclik, T., V. Hartenstein, et al. (2008). "Conserved role of the Vsx genes supports a monophyletic origin for bilaterian visual systems." Curr Biol **18**(17): 1278-87.

Filipowicz, W. (2005). "RNAi: the nuts and bolts of the RISC machine." Cell **122**(1): 17-20.

Fischer, J. A., E. Giniger, et al. (1988). "GAL4 activates transcription in Drosophila." Nature **332**(6167): 853-6.

Giniger, E., K. Tietje, et al. (1994). "lola encodes a putative transcription factor required for axon growth and guidance in Drosophila." Development **120**(6): 1385-98.

Giniger, E., S. M. Varnum, et al. (1985). "Specific DNA binding of GAL4, a positive regulatory protein of yeast." Cell **40**(4): 767-74.

Grimson, A., M. Srivastava, et al. (2008). "Early origins and evolution of microRNAs and Piwi-interacting RNAs in animals." Nature **455**(7217): 1193-7.

Gubb, D., C. Green, et al. (1999). "The balance between isoforms of the prickle LIM domain protein is critical for planar polarity in Drosophila imaginal discs." Genes Dev **13**(17): 2315-27.

Guyer, D., A. Tuttle, et al. (1998). "Activation of latent transgenes in Arabidopsis using a hybrid transcription factor." Genetics **149**(2): 633-9.

Haas, M. S., S. J. Brown, et al. (2001). "Homeotic evidence for the appendicular origin of the labrum in Tribolium castaneum." Dev Genes Evol **211**(2): 96-102.

Haas, M. S., S. J. Brown, et al. (2001). "Pondering the procephalon: the segmental origin of the labrum." Dev Genes Evol **211**(2): 89-95.

Handler, A. M. and R. A. Harrell, 2nd (1999). "Germline transformation of Drosophila melanogaster with the piggyBac transposon vector." Insect Mol Biol **8**(4): 449-57.

Hartley, K. O., S. L. Nutt, et al. (2002). "Targeted gene expression in transgenic Xenopus using the binary Gal4-UAS system." Proc Natl Acad Sci U S A **99**(3): 1377-82.

Horn, C., B. G. Schmid, et al. (2002). "Fluorescent transformation markers for insect transgenesis." Insect Biochem Mol Biol **32**(10): 1221-35.

Horn, C. and E. A. Wimmer (2000). "A versatile vector set for animal transgenesis." Dev Genes Evol **210**(12): 630-7.

Hu, S., D. Fambrough, et al. (1995). "The Drosophila abrupt gene encodes a BTB-zinc finger regulatory protein that controls the specificity of neuromuscular connections." Genes Dev **9**(23): 2936-48.

Hunt, T., J. Bergsten, et al. (2007). "A comprehensive phylogeny of beetles reveals the evolutionary origins of a superradiation." Science **318**(5858): 1913-6.

Imamura, M., J. Nakai, et al. (2003). "Targeted gene expression using the GAL4/UAS system in the silkworm Bombyx mori." Genetics **165**(3): 1329-40.

Johnston, C. I. (1992). "Franz Volhard Lecture. Renin-angiotensin system: a dual tissue and hormonal system for cardiovascular control." J Hypertens Suppl 10(7): S13-26.

Jürgens, G., Lehmann, R., Schardin, M., Nusslein-Volhard, C. (1986). "Segmental organisation of the head in the embryo of Drosophila melanogaster." Roux's Arch. Dev. Biol. 195: 359–377.

Juven-Gershon, T., S. Cheng, et al. (2006). "Rational design of a super core promoter that enhances gene expression." Nat Methods 3(11): 917-22.

Konopova, B. and M. Jindra (2007). "Juvenile hormone resistance gene Methoprene-tolerant controls entry into metamorphosis in the beetle Tribolium castaneum." Proc Natl Acad Sci U S A 104(25): 10488-93.

Konopova, B. and M. Jindra (2008). "Broad-Complex acts downstream of Met in juvenile hormone signaling to coordinate primitive holometabolan metamorphosis." Development 135(3): 559-68.

Laughon, A., R. Driscoll, et al. (1984). "Identification of two proteins encoded by the Saccharomyces cerevisiae GAL4 gene." Mol Cell Biol 4(2): 268-75.

Laughon, A. and R. F. Gesteland (1984). "Primary structure of the Saccharomyces cerevisiae GAL4 gene." Mol Cell Biol 4(2): 260-7.

Li, W., F. Wang, et al. (2004). "BTB/POZ-zinc finger protein abrupt suppresses dendritic branching in a neuronal subtype-specific and dosage-dependent manner." Neuron 43(6): 823-34.

Lis, J. T., J. A. Simon, et al. (1983). "New heat shock puffs and beta-galactosidase activity resulting from transformation of Drosophila with an hsp70-lacZ hybrid gene." Cell 35(2 Pt 1): 403-10.

Lorenzen, M. D., A. J. Berghammer, et al. (2003). "piggyBac-mediated germline transformation in the beetle Tribolium castaneum." Insect Mol Biol 12(5): 433-40.

Lorenzen, M. D., T. Kimzey, et al. (2007). "piggyBac-based insertional mutagenesis in Tribolium castaneum using donor/helper hybrids." Insect Mol Biol 16(3): 265-75.

Lynch, J. and C. Desplan (2003). "Evolution of development: beyond bicoid." Curr Biol 13(14): R557-9.

Ma, J. and M. Ptashne (1987). "The carboxy-terminal 30 amino acids of GAL4 are recognized by GAL80." Cell 50(1): 137-42.

Ma, J. and M. Ptashne (1987). "Deletion analysis of GAL4 defines two transcriptional activating segments." Cell 48(5): 847-53.

Ma, J. and M. Ptashne (1987). "A new class of yeast transcriptional activators." Cell 51(1): 113-9.

Maderspacher, F., G. Bucher, et al. (1998). "Pair-rule and gap gene mutants in the flour beetle Tribolium castaneum." Dev Genes Evol 208(10): 558-68.

McGinnis, W. and R. Krumlauf (1992). "Homeobox genes and axial patterning." Cell 68(2): 283-302.

McGuire, S. E., G. Roman, et al. (2004). "Gene expression systems in Drosophila: a synthesis of time and space." Trends Genet 20(8): 384-91.

McKnight, J. L., T. M. Kristie, et al. (1987). "Binding of the virion protein mediating alpha gene induction in herpes simplex virus 1-infected cells to its cis site requires cellular proteins." Proc Natl Acad Sci U S A 84(20): 7061-5.

McNeill, H., C. H. Yang, et al. (1997). "mirror encodes a novel PBX-class homeoprotein that functions in the definition of the dorsal-ventral border in the Drosophila eye." Genes Dev 11(8): 1073-82.

Miller, S. C., S. J. Brown, et al. (2008). "Larval RNAi in Drosophila?" Dev Genes Evol 218(9): 505-10.

Morrison, D. K., M. S. Murakami, et al. (2000). "Protein kinases and phosphatases in the Drosophila genome." J Cell Biol **150**(2): F57-62.
Nassif, C., A. Daniel, et al. (1998). "The role of morphogenetic cell death during Drosophila embryonic head development." Dev Biol **197**(2): 170-86.
O'Donnell, K. H. and P. C. Wensink (1994). "GAGA factor and TBF1 bind DNA elements that direct ubiquitous transcription of the Drosophila alpha 1-tubulin gene." Nucleic Acids Res **22**(22): 4712-8.
O'Hare, P. and C. R. Goding (1988). "Herpes simplex virus regulatory elements and the immunoglobulin octamer domain bind a common factor and are both targets for virion transactivation." Cell **52**(3): 435-45.
O'Hare, P., C. R. Goding, et al. (1988). "Direct combinatorial interaction between a herpes simplex virus regulatory protein and a cellular octamer-binding factor mediates specific induction of virus immediate-early gene expression." Embo J **7**(13): 4231-8.
Ochman, H., A. S. Gerber, et al. (1988). "Genetic applications of an inverse polymerase chain reaction." Genetics **120**(3): 621-3.
Ornitz, D. M., R. W. Moreadith, et al. (1991). "Binary system for regulating transgene expression in mice: targeting int-2 gene expression with yeast GAL4/UAS control elements." Proc Natl Acad Sci U S A **88**(3): 698-702.
Parthasarathy, R. and S. R. Palli (2009). "Molecular analysis of juvenile hormone analog action in controlling the metamorphosis of the red flour beetle, Tribolium castaneum." Arch Insect Biochem Physiol **70**(1): 57-70.
Patterson, G., R. N. Day, et al. (2001). "Fluorescent protein spectra." J Cell Sci **114**(Pt 5): 837-8.
Patterson, G. H., S. M. Knobel, et al. (1997). "Use of the green fluorescent protein and its mutants in quantitative fluorescence microscopy." Biophys J **73**(5): 2782-90.
Posnien, N. (2009). Function and Evolution of highly conserved head genes in the red flour beetle Tribolium castaneum. Department of Developmental Biology. Göttingen, Georg-August-Universität
Preston, C. M., M. C. Frame, et al. (1988). "A complex formed between cell components and an HSV structural polypeptide binds to a viral immediate early gene regulatory DNA sequence." Cell **52**(3): 425-34.
Ramos, D. M., F. Kamal, et al. (2006). "Temporal and spatial control of transgene expression using laser induction of the hsp70 promoter." BMC Dev Biol **6**: 55.
Rempel, J. G. (1975). "The Evolution of the insect head: the endless dispute." Quaestiones entomologicae **11**: 7-25.
Richards, S., R. A. Gibbs, et al. (2008). "The genome of the model beetle and pest Tribolium castaneum." Nature **452**(7190): 949-55.
Rogers, B. T. and T. C. Kaufman (1997). "Structure of the insect head in ontogeny and phylogeny: a view from Drosophila." Int Rev Cytol **174**: 1-84.
Rorth, P. (1998). "Gal4 in the Drosophila female germline." Mech Dev **78**(1-2): 113-8.
Sadowski, I., J. Ma, et al. (1988). "GAL4-VP16 is an unusually potent transcriptional activator." Nature **335**(6190): 563-4.
Scheer, N. and J. A. Campos-Ortega (1999). "Use of the Gal4-UAS technique for targeted gene expression in the zebrafish." Mech Dev **80**(2): 153-8.
Schinko, J. B. (2003). Erweiterung des Satzes an universellen Transformationsvektoren zur genetischen Manipulation von Invertebraten. Bayreuth, Department of Genetics.
Schinko, J. B., N. Kreuzer, et al. (2008). "Divergent functions of orthodenticle, empty spiracles and buttonhead in early head patterning of the beetle Tribolium castaneum (Coleoptera)." Dev Biol **317**(2): 600-13.

Schmidt-Ott, U. and G. M. Technau (1992). "Expression of en and wg in the embryonic head and brain of Drosophila indicates a refolded band of seven segment remnants." Development **116**(1): 111-25.
Scholtz, G. and G. D. Edgecombe (2006). "The evolution of arthropod heads: reconciling morphological, developmental and palaeontological evidence." Dev Genes Evol **216**(7-8): 395-415.
Schroder, R. (2003). "The genes orthodenticle and hunchback substitute for bicoid in the beetle Tribolium." Nature **422**(6932): 621-5.
Skwarek, L. C. and G. L. Boulianne (2009). "Great expectations for PIP: phosphoinositides as regulators of signaling during development and disease." Dev Cell **16**(1): 12-20.
Somma, M. P., F. Ceprani, et al. (2008). "Identification of Drosophila mitotic genes by combining co-expression analysis and RNA interference." PLoS Genet **4**(7): e1000126.
Stebbins, M. J., S. Urlinger, et al. (2001). "Tetracycline-inducible systems for Drosophila." Proc Natl Acad Sci U S A **98**(19): 10775-80.
Sugimura, K., D. Satoh, et al. (2004). "Development of morphological diversity of dendrites in Drosophila by the BTB-zinc finger protein abrupt." Neuron **43**(6): 809-22.
Sulston, I. A. and K. V. Anderson (1996). "Embryonic patterning mutants of Tribolium castaneum." Development **122**(3): 805-14.
Suzuki, Y., D. C. Squires, et al. (2009). "Larval leg integrity is maintained by Distal-less and is required for proper timing of metamorphosis in the flour beetle, Tribolium castaneum." Dev Biol **326**(1): 60-7.
Szuts, D. and M. Bienz (2000). "LexA chimeras reveal the function of Drosophila Fos as a context-dependent transcriptional activator." Proc Natl Acad Sci U S A **97**(10): 5351-6.
Tautz, D. and C. Pfeifle (1989). "A non-radioactive in situ hybridization method for the localization of specific RNAs in Drosophila embryos reveals translational control of the segmentation gene hunchback." Chromosoma **98**(2): 81-5.
Tomoyasu, Y. and R. E. Denell (2004). "Larval RNAi in Tribolium (Coleoptera) for analyzing adult development." Dev Genes Evol **214**(11): 575-8.
Tomoyasu, Y., S. C. Miller, et al. (2008). "Exploring systemic RNA interference in insects: a genome-wide survey for RNAi genes in Tribolium." Genome Biol **9**(1): R10.
Triezenberg, S. J., R. C. Kingsbury, et al. (1988). "Functional dissection of VP16, the trans-activator of herpes simplex virus immediate early gene expression." Genes Dev **2**(6): 718-29.
Triezenberg, S. J., K. L. LaMarco, et al. (1988). "Evidence of DNA: protein interactions that mediate HSV-1 immediate early gene activation by VP16." Genes Dev **2**(6): 730-42.
Tsukiyama, T., P. B. Becker, et al. (1994). "ATP-dependent nucleosome disruption at a heat-shock promoter mediated by binding of GAGA transcription factor." Nature **367**(6463): 525-32.
Viktorinova, I. and E. A. Wimmer (2007). "Comparative analysis of binary expression systems for directed gene expression in transgenic insects." Insect Biochem Mol Biol **37**(3): 246-54.
Vincent, A., J. T. Blankenship, et al. (1997). "Integration of the head and trunk segmentation systems controls cephalic furrow formation in Drosophila." Development **124**(19): 3747-54.

Yang, T. T., L. Cheng, et al. (1996). "Optimized codon usage and chromophore mutations provide enhanced sensitivity with the green fluorescent protein." Nucleic Acids Res **24**(22): 4592-3.

Yang, X., M. Weber, et al. (2009). "Probing the Drosophila retinal determination gene network in Tribolium (II): The Pax6 genes eyeless and twin of eyeless." Dev Biol.

Zhong, J. and B. Yedvobnick (2009). "Targeted gain-of-function screening in Drosophila using GAL4-UAS and random transposon insertions." Genet Res **91**(4): 243-58.

6 Appendix

Appendix

6.1 Supplementary Table

Table A1: Summary of head bristle defects of the rescreened lethal lines

No	bell row		vertex setae			vertex triplet				labrum quartet		gena triplet				maxilla escort			empty eggs
	bells	bristle	post	vent	ant	triplet	ant bas	median	ant	clypeus	labrum	post	dors	ant	bristle	ant	med	post	
G02408	80%	20%	10%	10%	20%	55%	10%	15%	20%	0%	0%	30%	0%	25%	30%	0%	0%	0%	61%
G07411	100%	100%	10%	0%	0%	30%	0%	10%	0%	0%	0%	0%	0%	0%	0%	0%	0%	0%	15%
G07521	Only very view cuticles that exhibit phenotype of reduced distance between ventral and anterior vertex setae. All other bristles not affected																		55%
G09104	0%	0%	0%	0%	0%	0%	0%	0%	0%	0%	0%	0%	0%	0%	0%	0%	0%	0%	60%
G10215	36%	7%	14%	0%	21%	14%	7%	7%	14%	0%	0%	21%	14%	0%	0%	0%	0%	0%	19%
KS0294	6%	0%	0%	6%	50%	19%	6%	38%	50%	88%	100%	0%	0%	0%	0%	0%	0%	0%	12%
KT1269	8%	33%	66%	83%	66%	50%	66%	50%	25%	8%	0%	0%	58%	0%	83%	25%	0%	25%	29%

6.2 Sequences of rescreen of lethal lines

G02408:

*piggy*Bac left arm flanking sequence:

TTAAGCATTGTTGATTTAGAAAAACACTATTGAAATGATAAAATTAATAATTAATTAATTACACGTAATAATATTTT
AAATTTGATTAATATAATTTTTTAGTAATTAATGTTTATTTAAGGAAATGTAGAAAAATAAAACAATCCGAAAAAA
GTTATTAATAGAAAACTCTGATTTCTGTGTGATTAATTAGTTTATGAGAAAATGTTAACTATACTTTTAGTAAACTC
TGTAGTTAAATTAATGTGCCAACATTCTTTAATAAATGATTTTGTCTTTTTATTAATTATTGTTGCATTATGAGTGA
TAACTAATTCTGTAAGTGGGGAAACAATTTGCCCCGGAAATCACCGGGAGATC

Glean_10926 DNA sequence (the cloned sequence is marked in yellow):

ATGCCTGTCTCTAGCGATGTTGTTCGTAAAGAAACCCCCATGGAAATTATATTTACAGACGAAGACCTCCCTTACGA
AGAAGAGATCCTCCGCAACCCCTACTCTGTGAAGCACTGGCTGCGCTACATTGAGCACAAGAAGAAGGCCCCGAAGC
ACGGCGTTAACATAATTTACGAGCGGGCCCTTAAAGAGCTCCCCGGCTCTTACAAACTTTGGTACAACTACTTACGC
ACTCGCCGCCTCCAGGTGAAAACCGCTGCATCACAGACCCCGCATTCGAGGAAGTCAACAACGCTTTTGAGCGCTC
CCTTGTTTTCATGCACAAAATGCCGCGAATTTGGATGGATTATTGCAGTTTCCTGACCGATCAGTGCAAGATAACGC
GGACTCGGAAGGTTTTCGACCGGGCTTTGAGGGCGCTCCCGGTGACCCAACACCACCGGATTTGGCCCCTTTACCTC
ACATTCGTCAAAAAGCACGACATTTCCGAGACCGCTGTGCGCATTTTTCGCCGGTATTTGAAGCTCAGTCCGGAGAA
TGCCGAGGAATACGTTGAGTATTTGACCGAAGTTGGGCGTTTGGACGAAGCCGCCGTTGTGTTGGCTAAGATTGTCA
ATGATGAGAATTTCGTGTCACAGCACGGGAAGTCCAAACACCAGCTCTGGAACGAGTTGTGCGAGTTGATTTCGAAA
AATCCCGAAGAGGTGCATTCGTTGAATGTTGATGCGATTATCAGGGCGGCTTGAGGCGGTACACTGATCAGTTGGG
GCATTTGTGGAACTCTCTGGCTACGTATTATGTCAGGAGCGGGTTGTTTGAACGGGCGCGGGATATTTACGAGGAGG
CGATTCAGACGGTTACAACTGTGCGGGATTTTACGCAAGTCTTTGACGCCTATGCACAGTTTGAGGAGTTGACGTTG
AGTAAACGGATGGAGGAGGTGGCGCAAAAGCCCAATCAGACTGAAGATGATGATATTGAGTTGGAGTTGAGGCTGGC
GAGGTTTGAGAATTTGATGGAGAGGAGGCTGTTGCTGTTGAATTCGGTTCTCTTGAGGCAGAATCCGCATAATGTGC
AGGAGTGGCACAAGCGAGTGCAGCTGTATGAAGGAAAACCACACGAAATCATCAACACCTACACCGAGGCGGTGCAA
ACCGTCGACCCCAAACTCGCCGTAGGCAAACTGCACACCCTCTGGGTCGAATTCGCCAAATTCTACGAAACCAACAA
ACAAATCCAGGACGCTCGTCTCATTTTCGAGAAAGCCACTCAAGTGGCGTACGTGAAAGTGGACGATTTGGCGACGG
TTTGGTGCGAATGGGCCGAAATGGAGATACGAAACGAAAACTACGAGCAAGCCCTCAAACTCATGCACAGAGCTAGC
ACCATGCCATCACGCAAAGTGCCTATCATGACGACACCGAGACCGTGCAAGCCAGACTGTACAAGTCGCTGAAAGT
CTGGTCGATGTTGGCCGACTTGGAGGAGAGTTTCGGGACGTTTAAGTCATGCAAAGCGGTCTACGACAGGATTATTG
ACTTGAAAATTGCAACCCCGCAGATTATTATCAATTACGGGTTGTTCCTAGAGGAGAATAATTATTTTGAGGAGGCG
TTCCGGGCCTATGAGAAGGGGATTTCGCTGTTCAAGTGGCCGAACGTTTACGACATTTGGAACACGTACTTGAGCAA
GTTCTTGAAGCGTTATGGGGGAGCAAGTTGGAGAGGGCGCGCGATTTGTTCGAGCAGTGTTTGGAGAACTGTCCTC
CACAATTCGCCAAACCCCTCTACCTCCTTTACGCGAAGTTGGAGGAGGAGCACGGAATGGCCCGACACGCCATGGCT
GTGTACGAAAGGGCCACAAACGCTGTCCCTCAGGAGGAAATGTTCGAAATATTCAACATTTACATCAAAAGGGCGGC
CGAAATTTACGGGATTCCCAAAACGCCCAGATTTACGAGAAAGCGATTGAGGTCCTTCCGGAGGACAAAACTCGGG
AGATGTGCGTCCGTTTCGCCGATATGGAGACGAAGTTGGGCGAGATCGACCGAGCTCGCGCCATCTACTCGCACTGT
AGCCAGATTTGCGACCCGAGGGTCACGACTGAGTTCTGGCAGATCTGGAAGGAGTTTGAGGTGAGGCACGGGAATGA
GGACACGATGAGAGAGATGTTGAGGATCAAGAGGAGTATCCAAGCGATGTACAACACTCAGATTAATATGATGTCGG
CTCAGATGCTGAGTTCGGCGAGTTCCGTGGCGGGGACTGTTGCTGATCTGGCGCCGGGGGTTAAGGATGGGATGCGG
TTGCTTGAGGCGAAGGCGGCCGAAATGGCAGGGACGTCGAGTGGGGTCCGGCCTGGGAATATTATGTTTGTGAGGGG
AGAGACGCAAGGGGATAATAAGGATAAGGTTGTTAATCCGGATGAGATTGATATTGGGGATGATGACGACGAGGAGG
AGAATGAGGACGAAGGGGAGGATGTGCCGGTTGAGAAGCAGAGTATTCCGTCGGAAGTGTTTGGTGGGTTGAAGAAG
AAGGATGATCAGGGGGAACAGGATGAGGGATGA

Appendix

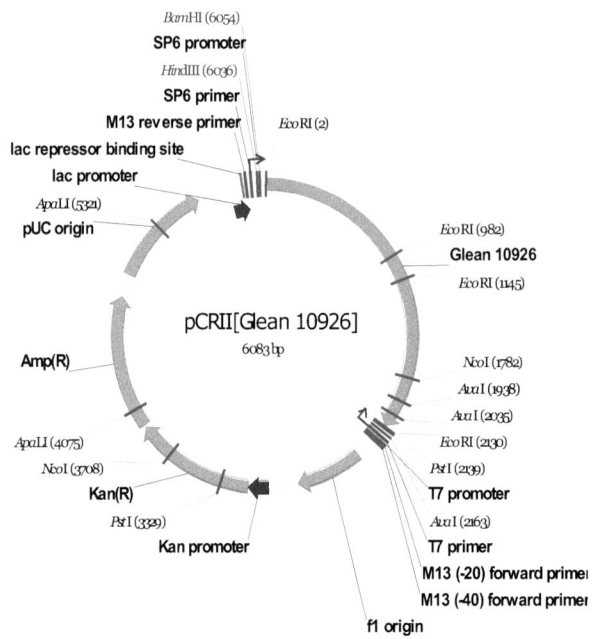

>pCRII[Glean_10926]

```
GAATTCGGCTTTTACGAAGAAGAGATCCTCCGCAACCCCTACTCTGTGAAGCACTGGCTGCGCTACATTGAGCACAA
GAAGAAGGCCCCGAAGCACGGCGTTAACATAATTTACGAGCGGGCCCTTAAAGAGCTCCCCGGCTCTTACAAACTTT
GGTACAACTACTTACGCACTCGCCGCCTCCAGGTGAAAAACCGCTGCATCACAGACCCCGCATTCGAGGAAGTCAAC
AACGCTTTTGAGCGCTCCCTTGTTTTCATGCACAAAATGCCGCGAATTTGGATGGATTATTGCAGTTTCCTGACCGA
TCAGTGCAAGATAACGCGGACTCGGAAGGTTTTCGACCGGGCTTTGAGGGCGCTCCCGGTGACCAACACCACCGGA
TTTGGCCCCTTTACCTCACATTCGTCAAAAAGCACGACATTTCCGAGACCGCTGTGCGCATTTTTCGCCGGTATTTG
AAGCTCAGTCCGGAGAATGCGGAGGAATACGTTGAGTATTTGACCGAAGTTGGGCGTTTGGACGAAGCCGCCGTTGT
GTTGGCTAAGATTGTCAATGATGAGAATTTCGTGTCACAGCACGGGAAGTCCAAACACCAGCTCTGGAACGAGTTGT
GCGAGTTGATTTCGAAAAATCCCGAAGAGGTGCATTCGTTGAATGTTGATGCGATTATCAGGGGCGGCTTGAGGCGG
TACACTGATCAGTTGGGGCATTTGTGGAACTCTCTGGCTACGTATTTATGTCAGGAGCGGGTTGTTTGAACGGGCGCG
GGATATTTACGAGGAGGCGATTCAGACGGTTACAACTGTGCGGGATTTTACGCAAGTCTTTGACGCCTATGCACAGT
TTGAGGAGTTGACGTTGAGTAAACGGATGGAGGAGGTGGCGCAAAAGCCCAATCAGACTGAAGATGATGATATTGAG
TTGGAGTTGAGGCTGGCGAGGTTTGAGAATTTGATGGAGAGGAGGCTGTTGCTGTTGAATTCGGTTCTCTTGAGGCA
GAATCCGCATAATGTGCAGGAGTGGCACAAGCGAGTGCAGCTGTATGAAGGAAAACCACACGAAATCATCAACACCT
ACACCGAGGCGGTGCAAACCGTCGACCCCAAACTCGCCGTAGGCAAACTGCACACCCTCTGGGTCGAATTCGCCAAA
TTCTACGAAACCAACAAACAAATCGAGGACGCTCGTCTCATTTTCGAGAAAGCCACTCAAGTGGCGTACGTGAAAGT
GGACGATTTGGCGACGGTTTGGTGCGAATGGGCCGAAATGGAGATACGAAACGAAAACTACGAGCAAGCCCTCAAAC
TCATGCACAGAGCTAGCACCATGCCATCACGCAAAGTGGCCTATCATGACGACACCGAGACCGTGCAAGCCAGACTG
TACAAGTCGCTGAAAGTCTGGTCGATGTTGGCCGACTTGGAGGAGAGTTTCGGGACGTTTAAGTCATGCAAAGCGGT
CTACGACAGGATTATTGACTTGAAAATTGCAACCCCGCAGATTATTATCAATTACGGGTTGTTCCTAGAGGAGAATA
ATTATTTTGAGGAGGCGTTCCGGGCCTATGAGAAGGGGATTTCGCTGTTCAAGTGGCCGAACGTTTACGACATTTGG
AACACGTACTTGAGCAAGTTCTTGAAGCGTTATGGGGGGAGCAAGTTGGAGAGGGCGCGCGATTTGTTCGAGCAGTG
TTTGGAGAACTGTCCTCCACAATTCGCCAAACCCCTCTACCTCCTTTACGCGAAGTTGGAGGAGGAGCACGGAATGG
CCCGACACGCCATGGCTGTGTACGAAAGGGCCACAAACGCTGTCCCTCAGGAGGAAATGTTCGAAATATTCAACATT
TACATCAAAAGGGCGGCCGAAATTTACGGGATTCCCAAAACGCGCCAGATTTACGAGAAAGCGATTGAGGTCCTTCC
GGAGGACAAAACTCGGGAGATGTGCGTCCGTTTCGCCGATATGGAGACGAAGTTGGGCGAGATCGACCGAGCTCGCG
CCATCTACTCGCACTGTAGCCAGATTTGCGACCCGAGGGTCACGACTGAGTTCTGGCAGATCTGGAAGGAGTTTGAG
GTGAGGCACGGGAATGAGGACACGATGAGAGAGATGTTGAGGATAAGCCGAATTCTGCAGATATCCATCACACTGGC
GGCCGCTCGAGCATGCATCTAGAGGGCCCAATTCGCCCTATAGTGAGTCGTATTACAATTCACTGGCCGTCGTTTTA
```

Appendix

```
CAACGTCGTGACTGGGAAAACCCTGGCGTTACCCAACTTAATCGCCTTGCAGCACATCCCCCTTTCGCCAGCTGGCG
TAATAGCGAAGAGGCCCGCACCGATCGCCCTTCCCAACAGTTGCGCAGCCTGAATGGCGAATGGACGCGCCCTGTAG
CGGCGCATTAAGCGCGGCGGGTGTGGTGGTTACGCGCAGCGTGACCGCTACACTTGCCAGCGCCCTAGCGCCCGCTC
CTTTCGCTTTCTTCCCTTCCTTTCTCGCCACGTTCGCCGGCTTTCCCCGTCAAGCTCTAAATCGGGGGCTCCCTTTA
GGGTTCCGATTTAGTGCTTTACGGCACCTCGACCCCAAAAAACTTGATTAGGGTGATGGTTCACGTAGTGGGCCATC
GCCCTGATAGACGGTTTTTCGCCCTTTGACGTTGGAGTCCACGTTCTTTAATAGTGGACTCTTGTTCCAAACTGGAA
CAACACTCAACCCTATCTCGGTCTATTCTTTTGATTTATAAGGGATTTTGCCGATTTCGGCCTATTGGTTAAAAAAT
GAGCTGATTTAACAAAAATTTAACGCGAATTTTAACAAAATTCAGGGCGCAAGGGCTGCTAAAGGAAGCGGAACACG
TAGAAAGCCAGTCCGCAGAAACGGTGCTGACCCCGGATGAATGTCAGCTACTGGGCTATCTGGACAAGGGAAAACGC
AAGCGCAAAGAGAAAGCAGGTAGCTTGCAGTGGGCTTACATGGCGATAGCTAGACTGGGCGGTTTTATGGACAGCAA
GCGAACCGGAATTGCCAGCTGGGGCGCCCTCTGGTAAGGTTGGGAAGCCCTGCAAAGTAAACTGGATGGCTTTCTTG
CCGCCAAGGATCTGATGGCGCAGGGGATCAAGATCTGATCAAGAGACAGGATGAGGATCGTTTCGCATGATTGAACA
AGATGGATTGCACGCAGGTTCTCCGGCCGCTTGGGTGGAGAGGCTATTCGGCTATGACTGGGCACAACAGACAATCG
GCTGCTCTGATGCCGCCGTGTTCCGGCTGTCAGCGCAGGGGCGCCCGGTTCTTTTTGTCAAGACCGACCTGTCCGGT
GCCCTGAATGAACTGCAGGACGAGGCAGCGCGGCTATCGTGGCTGGCCACGACGGGCGTTCCTTGCGCAGCTGTGCT
CGACGTTGTCACTGAAGCGGGAAGGGACTGGCTGCTATTGGGCGAAGTGCCGGGGCAGGATCTCCTGTCATCCCACC
TTGCTCCTGCCGAGAAAGTATCCATCATGGCTGATGCAATGCGGCGGCTGCATACGCTTGATCCGGCTACCTGCCCA
TTCGACCACCAAGCGAAACATCGCATCGAGCGAGCACGTACTCGGATGGAAGCCGGTCTTGTCGATCAGGATGATCT
GGACGAAGAGCATCAGGGGCTCGCGCCAGCCGAACTGTTCGCCAGGCTCAAGGCGCGCATGCCCGACGGCGAGGATC
TCGTCGTGACCCATGGCGATGCCTGCTTGCCGAATATCATGGTGGAAAATGGCCGCTTTTCTGGATTCATCGACTGT
GGCCGGCTGGGTGTGGCGGACCGCTATCAGGACATAGCGTTGGCTACCCGTGATATTGCTGAAGAGCTTGGCGGCGA
ATGGGCTGACCGCTTCCTCGTGCTTTACGGTATCGCCGCTCCCGATTCGCAGCGCATCGCCTTCTATCGCCTTCTTG
ACGAGTTCTTCTGAATTGAAAAAGGAAGAGTATGAGTATTCAACATTTCCGTGTCGCCCTTATTCCCTTTTTTGCGG
CATTTTGCCTTCCTGTTTTTGCTCACCCAGAAACGCTGGTGAAAGTAAAAGATGCTGAAGATCAGTTGGGTGCACGA
GTGGGTTACATCGAACTGGATCTCAACAGCGGTAAGATCCTTGAGAGTTTTCGCCCCGAAGAACGTTTTCCAATGAT
GAGCACTTTTAAAGTTCTGCTATGTGGCGCGGTATTATCCCGTATTGACGCCGGGCAAGAGCAACTCGGTCGCCGCA
TACACTATTCTCAGAATGACTTGGTTGAGTACTCACCAGTCACAGAAAAGCATCTTACGGATGGCATGACAGTAAGA
GAATTATGCAGTGCTGCCATAACCATGAGTGATAACACTGCGGCCAACTTACTTCTGACAACGATCGGAGGACCGAA
GGAGCTAACCGCTTTTTTGCACAACATGGGGGATCATGTAACTCGCCTTGATCGTTGGGAACCGGAGCTGAATGAAG
CCATACCAAACGACGAGCGTGACACCACGATGCCTGTAGCAATGGCAACAACGTTGCGCAAACTATTAACTGGCGAA
CTACTTACTCTAGCTTCCCGGCAACAATTAATAGACTGGATGGAGGCGGATAAAGTTGCAGGACCACTTCTGCGCTC
GGCCCTTCCGGCTGGCTGGTTTATTGCTGATAAATCTGGAGCCGGTGAGCGTGGGTCTCGCGGTATCATTGCAGCAC
TGGGGCCAGATGGTAAGCCCTCCCGTATCGTAGTTATCTACACGACGGGGAGTCAGGCAACTATGATGAACGAAAT
AGACAGATCGCTGAGATAGGTGCCTCACTGATTAAGCATTGGTAACTGTCAGACCAAGTTTACTCATATATACTTTA
GATTGATTTAAAACTTCATTTTTAATTTAAAAGGATCTAGGTGAAGATCCTTTTTGATAATCTCATGACCAAAATCC
CTTAACGTGAGTTTTCGTTCCACTGAGCGTCAGACCCCGTAGAAAAGATCAAAGGATCTTCTTGAGATCCTTTTTTT
CTGCGCGTAATCTGCTGCTTGCAAACAAAAAAACCACCGCTACCAGCGGTGGTTTGTTTGCCGGATCAAGAGCTACC
AACTCTTTTTCCGAAGGTAACTGGCTTCAGCAGAGCGCAGATACCAAATACTGTTCTTCTAGTGTAGCCGTAGTTAG
GCCACCACTTCAAGAACTCTGTAGCACCGCCTACATACCTCGCTCTGCTAATCCTGTTACCAGTGGCTGCTGCCAGT
GGCGATAAGTCGTGTCTTACCGGGTTGGACTCAAGACGATAGTTACCGGATAAGGCGCAGCGGTCGGGCTGAACGGG
GGGTTCGTGCACACAGCCCAGCTTGGAGCGAACGACCTACACCGAACTGAGATACCTACAGCGTGAGCTATGAGAAA
GCGCCACGCTTCCCGAAGGGAGAAAGGCGGACAGGTATCCGGTAAGCGGCAGGGTCGGAACAGGAGAGCGCACGAGG
GAGCTTCCAGGGGAAACGCCTGGTATCTTTATAGTCCTGTCGGGTTTCGCCACCTCTGACTTGAGCGTCGATTTTT
GTGATGCTCGTCAGGGGGCGGAGCCTATGGAAAAACGCCAGCAACGCGGCCTTTTTACGGTTCCTGGCCTTTTGCT
GGCCTTTTGCTCACATGTTCTTTCCTGCGTTATCCCCTGATTCTGTGGATAACCGTATTACCGCCTTTGAGTGAGCT
GATACCGCTCGCCGCAGCCGAACGACCGAGCGCAGCGAGTCAGTGAGCGAGGAAGCGGAAGAGCGCCCAATACGCAA
ACCGCCTCTCCCCGCGCGTTGGCCGATTCATTAATGCAGCTGGCACGACAGGTTTCCCGACTGGAAAGCGGGCAGTG
AGCGCAACGCAATTAATGTGAGTTAGCTCACTCATTAGGCACCCCAGGCTTTACACTTTATGCTTCCGGCTCGTATG
TTGTGTGGAATTGTGAGCGGATAACAATTTCACACAGGAAACAGCTATGACCATGATTACGCCAAGCTATTTAGGTG
ACACTATAGAATACTCAAGCTATGCATCAAGCTTGGTACCGAGCTCGGATCCACTAGTAACGGCCGCCAGTGTGCTG
```

G07411

*piggy*Bac right arm flanking sequence:

```
TTAAGTTACATTAAGAGCAGCATTTGTTAAAAAATTCCCACCCCTAAAGTTAGTAGTGAAAAGTAAGAAAGTTTTTT
AAAATTTTATTGTAGTAATAAATAACAAATTAAATCAACATAAAAATACATTCTAACAAATTTACAAAATATCTTAA
AGAAACAGTTATGGCAAATATAATAATTAGTTTTATTAGTGCAAGTGTTATGACCACGATGGCACTCAAGACGACGC
TGTACCTGCTGCTCATTTGGTCTATGGTTGAAAAACGATTCTAAACTAAACTGAAATATTCTTGGTGGTCTTCCTCA
GAAAAGTCGCTTTCACCTTTCTTGTTGTTTTTCTTTAGTATATTTTTTCCCTTTTTTTTACTACATTAATTTTTTT
TAATTGTATATTTATTGTCTTGAAATAACTGGTTTATATTCCCCGTTCTTCTTCTCCTACTTCTAAATTCGTTGGTG
TGTATATGTTATAACCCCAATCTCTCTGCCTCTTTTGTTACCTTTTGCATCCCTCTCCCTTCCCGCACAGTCAAAGC
AATATTCCCT
```

GLEAN_03634 DNA sequence:

```
ATGGCAGTAACCTCTTCGGCTCCAACTCCGGGAAGTCCGCCCGCTCCGTCTCAAAGATGCTGTGATACCGGCAGACC
TCTTTTCACCGACCCGCTAACCGGTCAAAGTGTGTGCTCGTGTCAGTATGAACTTTTAGGATATCAAAGATTAGCCG
GCGGAGTCCCGGGCCTCCCGGCCCTCTCCATGTACAGCGCGCCCTATCCCGAAGGAATGGCGGCCTACTTCCCAGCG
CTGGGGGCCGACCAAGCCCCCTTCTACACTTCGACGGCTGCCGGTTTAGAATTGAAAGAAAACTTGGCAGCGGGGGC
TGCGACATGGCCCTACCCTTCCGTTTATCATCCCTATGATACCGCCTTCGCAGGATATCCATTTAATGGATATGGGA
TGGATCTCAATGGTGCGCGTCGAAAAAACGCCACCAGAGAAACGACCAGTACACTCAAGGCATGGCTTAACGAACAC
AAAAAGAACCCGTATCCAACCAAAGGCGAAAAATTATGTTGGCCATTATTACAAAAATGACCTTGACGCAGGTCTC
GACGTGGTTCGCGAACGCGCGGAGGCGACTCAAGAAGGAGAACAAGATGACGTGGGAGCCGCGAAACAGAGTCGAAG
ATGAGGACAACAACAACGACGATGACGACCACAAGAGCACAGACGGCAAAGACATTCTAGATTCTAAAGATTCGGGC
ACCGCGTCCAGCGAAGACGGCGACCGACCACCGCATTCGCGTCTCGCTCCGGACACAGCGAGCGAATGGAGCGAATC
GCGCCCCGATAGCGGCCCGGACAGCCCCGAGTGCTACGAGCGCCCCCCTCACCCCGCCTTCCTGCCGCCGCGCTCGT
CCGGCTCTCCGCCGCAGATCTCGGCGTCCGTGAGCTCCAAGCCCCGGATTTGGTCGTTGGCCGACATGGCCAGCAAG
GAGAGCGACGGTCCGCCGCCGTCTACCGCCTCCTCGCTATACTCCACGGCGGCGAGCCGCCTCCGCTTGCCGCCCCC
GGGCCTCCACACCGCCCCCTACGCCCGCCCGCACGACTTCTACCGGAGTCTGTACGGGCCTGGGATGGCCGGCAGCG
GCTCTCCCGACGCCTCTCTCCTCGAAACGTACTCCAGGACCCTCGCCGGGCAGTCGGTGCTGACGAAGGCGAATCTG
GCGACGGCCACGTCCATTGCGAGCAATGGCCCCTTGGGGCTGACCACGAGTTCGCGGCTGTCGCCGTCGTCGACGTC
GTCGCTGTCGTCCGGCTCGGAGACGCCCTTGAAGCCGTCTCTAGAGGCGCTGGACCGTTCCGGCGCCTTAGTTCGTT
CTCTTGTAAATAAAGTGACTGCCATCGGCAGCAGTATAAAAACTTGTATCTATATTGTTTTATTTAGCCATCCCGTC
TGTTTGTGCGGTCCCCGGCATCAAGTTCCGAATGCGCTGACAATAGAGGTATGTCCCCGGTGCAAATCTCCGCCGGA
CCACCAGTCGAGGACAGCGATTTGGAGCGGAGTTGCAATGCATCAAACCGTGCACGAAATTAAGAGTCCACGAATGT
TTGACAAAACCGAAATTATTATGCCAGACACGAGTGGAGTTTTTCCAAGCGCGCGCACTTGA
```

G07521:

Insertion 1:

piggyBac right arm flanking sequence:

TTAAAGCGTTTCTCATTTTGTGCCGCAATCCATTATAATCTAGATCGAACCGTTTCGACGTCTCCCCCAGCATAAAA
AATTATAACACGTGTAATGTTTGAAAGTTTGATAAGGAAAAGTTAATCGGGTTTTTTTCAAGCCCATTGACGCGTCG
TAATCCAGCATTTTCAGTAAAAGATAAAAGAATGGTTGACGTTAGTTACTCGGCCACGTGGAGAGAACCGCAATAAA
CTCTCAAATAAAAATTCCACTGTCTTCCGAACGATTTGATTTCTAATAGGGTTAGCACTTTATGAGGTGCTTCACTA
AACAATTCCTGCGC

>GLEAN_00260

ATGAAATCGGGAGAAAGGAAGGTGGGAGATGCGCCTGCTCCATGGAGGGCGCTGGACCACCAACGCAACCTGCGGCC
TGCTTCAGTTCGTGCTCTTTTACGCGGCTTTTCGGTCGTGGTCCTTGACGAGGTCCTCGAAACGAGATCAGTGAGCC
TAGCAGGAACTGTTGATTCGGGCGTTTTCGCTCACCCGGACTCTGGCACCAGTTCGATTGGGCTTTCGTCGGCGATA
AATTGGAGGTCTGTAGCGCCAACGCCACATTTTACGACATATTTTACGGTCCGGGGCTATTTCATTTTTAATTCTAG
GGGGCGTGTTGGATGTCGTCACGCCGGCAGCGAATGCCCCGTAACTCATTCGACCCCCCAATGTGCTAGGAAACCTG
GCGTTCGTTGCGATGTCATCGTTGCCGCCAAAACACGAAATCCAAACTCAACCACCCCTTTCGATCGTAACGTTGAA
TATGTAAAGGTTTTGCCATTAAGTTTTAAACAAGCAAGTTATGCATATTTATCGCCGGCGAAATTTAATTTACGCGC
GCCCCCTTGCCGACCATTAGACAACTTTTAA

Insertion 2:

piggyBac left arm flanking sequence:

TTAAGAAAAATTCAGTTCTTTTGCAAAATTTTATTTCAGCGAACAGAAAAGTTACTACAATGGGTCAAAAAAACACC
GAGTGGTGATTTTATTTCGAGCAATTTCGTCAAAAATAAGCTGAAAAAAATCACTAAAAATACAAAAAGTGCTCTAA
GTTTCAAGAGTTTACGTACGTTTTTGGGCAAAACTATTTTTTTCTGAATACGACCAGTAACGCAAGAGATTATTTCA
CAAAAATTCGTCAAAAATAACTGAAACAATTACCGCGCTTTTTAGCTACGATGTCAAAGTTCGTAATTTTTGTGTTT
TGAGGTGGAAATTAATTTTTTTTATAAAATATTTTTAAACGAAGCTGAAAAAACTGAAAACTGACCTGAAATTATTA
TTTTTTGGGTATTTCCGGACGTTTTTCCTTAGGTTTTAAAAGGTTTTTTGTGAAGT

>GLEAN_07654

ATGGACCAGATTCACAGCTCACTAATTTTGTGGTGTGATCCTTTGTGTTTCCCCCACCGAAAGCTCGGCTTGCCAGC
GATCAGGTTACTCCCATTTCTTCCCAATTGTATAATTATCGACCGATTCCGCGTTAGAGAAAAAGAAATTTTTTCG
TAATAACTAATTCATCGGTAAACACCATCTGGATTGAAGAGGATGATTCAATTAAAAAGTATGATGCAGTGTTTCGA
GGGCCACCCCGTAGCATCAGACAGCAGCAGCAGAGCATCCAATTATTTTACAACGAACGGAATCGGCATCCAAATGC
CACTTTCACCAAAATCGAAAATCTTGCGAGAACTATTCGAAAACATGGACAACTAGACAATTCCAAAGACTTCTCAG
TGACTGACGGTTTAGGAAGCTTTAATAAAAAGAAGAAAAAGAAGAGGCGTCATAGGACGATTTTCACGAGCTACCAA
CTGGAAGAACTGGAGAAGGCCTTCAAAGATGCACACTATCCGGACGTTTACGCCAGGGAAATGCTTAGCCTGAAAAC
GGATTTACCAGAAGATCGGATACAGGTGTGGTTTCAGAACAGGAGGGCGAAATGGAGAAAAACTGAAAAATGTTGGG
GGAGGAGCAACCATCATGGCCGAATATGGCCTATATGGGGCTATGGTTCGGCATTCGCTACCTTTACCTGAAACAATT
CTAAAAAGCGCCAAAGAAAATGAATGTGTCGCGCCCTGGCTGTTAGGAATGCACCGAAAATCCTTGGAAGCCGCCCA
GCACCTGAAAGACCAAGACTGCACCAACAGCGACCATGAAGAAACGACTTCCATGCGAACGGAAGCAAGTGATACCA
GCGCCACCTCAAGCCAACAAACTCAACAACAGCAGCAGCCACCTCCCGGTCACACCGTTTCCGGTGGAGACGACGAT
CAACAGCAACAACACGTCGCTTTAGCAACAACACAACCACCACCCTATCAAGACGATCCGGAAGCCTTCAGAAACAA
CTCCATTGCCTGTTTGCGGGCTAAAGCTCAGGAGCACAGCGCCAAATTGCTCAATCATAATCTTATGATGCAGGTGA
GGTCTAGTGACGCCAACGCCAATAATTTACATCACCAGGAGGTGGTCAATGCAGAGACTAGTCCTAGTATTTTTGA

G09104:

*piggy*Bac left arm flanking sequence:

TTAAACACTTGATAAAACGTAAAAATAATGTCGTTCAGCTAAATTTTAAAGAACAAAGTGAATTGCTTTAATAAAAA
AACCTTCAAAAATACAGCATGTCTCAAAATTGGCGAATTCCTAGTCCAGAGCCTTGTAGAGAATCCCTTCAGTAGTC
CAAATATTTGAGACTCTCTGTAGAAAAAATACTTTGATTTTTTTGTGTTTGTAGTTAAAAAATATTTTTTAATTCAA
AAACTTGTTTAAATGCTTGAAACAACAACAATTAACACTTAACGACTTTTTTAATTATGACTTGTTTTAAATATAAA
ATTTTGTAGGAAAAAGTCAATTTTCTCAAAAACGTATCTATTCAAAGTTAAACGTTTGGGAAACCAGAAATAAAATA
ATTTAAAACAAATTTTCCGTTTTCGTTTATATTGTGTGTCTCTCTGATTAAAAAAAAAATAAAAAAACGGAAATGCT
GATGTGTTTCATCTTTTTTTTTTCCCCCAAAATTTTTAAGGCTAAGAAAATCGAATTTTTCCTGGCCAAAACCCTCC
CAAAAAAGTTGGTCTTCAAAA

GLEAN_14935 DNA sequence (the cloned sequence is marked in yellow):

ATGGCTAGAGATGATAAGCTTTTTTTCCACAAAACTGTGCAACATTTAGCCAGATCTCTAGCCTGCATTAAAAACAC
ACCATGGGATAAGGTGAAGACCTTGTACGACCTGTGCCCCATTGAGACAACCAACGGCACTGTCTCGTTAAACAGTC
GGCAGCAAGATGCCGTTATCGCCCTCGGGATCTACTTCCTGGAATCGGGACTCGAGCACAAGGACACGATTTTGCCC
TATTTACTCAAGCTGGCAAAAGCGCTTGGGAAAGCCACTTGGTTGGACGAAATTAAGCAGAATCCGAGCGATAGGAT
TCCCGTGGCTGAAAAGTTCAGTTTTTGTCTCCACACCCTTCTGTGTGACATCGCAGTGAAATGTGAAGATTCCCGGG
AAGAAATCATCGAAACACAAGTCGATTGTTTGGTGAAACTCACCAACAGTATCCTCAAAACTCGCGAAAATACAAAC
AGTGCAGTCAAACTCTTCCTGTGTAAAACCACAGTCCCTGTTTTAATTGGCCTGTCTCGGGCAATGGGGCGCTTCTG
CAACACTGAGCCCCCTTTAATCTGCCGCCTGTTCCCCAAACCGGAGCCCCCTTTATCCCCAGTCACCTCCAACCCCG
ACTATAAGCGCAGTTTTAGCAATTTCCGCTCAATCATTCCGAGAAGTCTGTCCGGTAATTTGGCTGCAACTGTTGAT
ATTTTGGCAATAACTCAAGGGTACGACACTACAGATGTTGCATATAGTTCTGCCAGTCTTAAAAGGGGCTCACTTAT
CAATCAAAATTTCGTCAATTATGATCCTGCAACTTATTTTTTCTCGAAATTCGGGTCCAGTTTTAATCAGTTTCCGC
ATTTACGTGTCAATGATCCGAATGATAAGAAGGGACAGATTATTTTTCCCTTGCAACATTTGCAAACGATACTTTCG
TTGGCGAAAAAACTGCTCACGAAAGATATGTTGAGTTTTTTGGATGAGCAAAGTTTGGAGGTTTATACGACGGGGAA
GATTGTTATTTTTCCGTATAAAACCTTCTCGGAGACGATTAATTTGGTTATGGTGACTTTGATGAGGGAGTTGCTTC
AGCCGCAGAAGAGTCTCCCGGTTGCGTTCACGAAGGATGTGCAAGAATTCGTTAAAGGGTTGTACTTAAACGGCCAG
ACGGAACTTCAGAGTCGCAACCACGACGCTAGTGAGAAAGAAGACCGTGAAAGTAATTTTGCCCTCGTTAATAAATT
CAAAGTTAACGTGATGGCAAACGCCGCCTGTGTTGATCTTCTAGTCTGGGCAATAGGGGACGAAACAGGTGCCGATA
GTCTTTGTGGCCGTCTTACCGAAAAAATCAACTCAAATCACGGCTACAAACTCGTCTTAGCCCACATGCCCCTCGTT
ATGGTGTGTCTTGAAGGCTTGGGCACTTTAGCCCAAAAATTCCCCAACATTGCCTCCACTTCCATTTACTGTCTTCG
AGATTTCCTCATCACTCCTTCACCCATCCTTTCCAAACTCCACCGACAAGCGAACGAAAAAGGAAACAAGGAAAATT
TAAAAATCATTGCCCACGTTAACGGCCTTAAGGCCGAAACCACTAAACCACACAACCCAACTCAATCCGCATTTGAG
AAACTACGAGACGCTGCTATTGAAAATCTATGTTATGCTCTTGAGGCTGCACACACTACTGATGCCGACTGTGTGAG
GGCGCTAGTGGCGTCGTTTCAAATCGGTTGTTTCGCGCTGAAAAAAGTGACAGTGAATCGACTCTCATCTCTACTA
ATATTGTTATAATGTTGGGTCATGTGGCCGTTTCGATGAAAGGCACCCCAAAGACAACGGATACGATTTTGCAGTTT
TTCCAACAGAGATTTTGTCGAGTGCCTTCGGCACTTGATACACTTATTGTTGACCAGTTGGGTTGTATGATTATCGC
ACAGTGTGAACCGCATGTTTACGAAGTTGTAATGAAGATGTTTACGATGATTACGGTGGAGAGTAGCAATGCGGCGT
ATGGGAATCCCACCAATGAGAAGGCACAGTATAGGCACGTTTCTCGCCCTGTGATCAACGCATTGGCCAACATCGCC
GCCAATATACAAAGCGAATCCCAAATGAACGAACTATTAGGCCGCCTGCTCGAACTCTTCGTGCAATTGGGCCTCGA
GGGCAAACGCGCAAGCGAAAAATCCCCAGGCGCCCTAAAAGCGTCAAGTTCAGCCGGTAATTTAGGAGTTTTAATCC
CTGTAATTGCCGTTCTTCTACGCCGACTACCACCCATCAAAAATCCCAAACCTCGGATTCATAAATTGTTCCGCGAT
TTTTGGCTCTATTGTGTTATTATGGGATTTACAGCGTCTGATTCGGGATTATGGCCGAAAGAATGGTACGAAGGAGT
GAAAGAAATCGCAGTTAAAAGCCCAGCGTTAGTTTCACCAACTTCCAGTCGCTCGGAAATGCGCGAATTGCAATACA
CGAGCGCCGTTCGAAACGATAGTGTTTCAATAACCGAACTACAAGAATTGAAGAATCAAATTCTAGAATTGTTGAAA
CAACCGGCAGATGTCACCGCCTATGTGAATCGCTTGACTTTTGCCCAATGTACGTTCTTATTGAGCGTTTATTGGGT
GGAGATTTTGCGAATTCAGAATTCACCTGAGCCGAGCTTGGTGCCGATTATTACCGAATACTTGTCGGATTCAGCGT
TACAGAAAGACAAGTCCGGGATGTGGGTTTGTGTGTCGGCTGTTGGCGAGCGCGTTTTTGAAAAATTCCTCGAAGTT
ATGAAAAATAAGCCGAAGTATGAGGCCAGGGAGGCGGAGCTTGAAGGCCACGCCCAGTTTCTTCTCGTACATTTCAA
CGATCCGCACAAACAGATTCGTCACGTTTCGGATAAGTTTCTGGTCAGTCTTTTCGATCGCTTTCCGCACTTGTTGT
GGGAGTCGGAAGGTGTTGTGGACAATGCTTGACATAATGCAAGTGTTGTCGAATTCCTTGCATTTGGATCCGAATCAG
GAAACGCCCACTTTGCGAATACCACGAACGCCATCTCCATACAGTTGATGACACGTTAGAGGCACGTAGAGACTAA
AGTTAAGCATTTTGCTGCGAATTCCGAGCGAATTATCAAGGAGGCGTTGAAGTGGGCACCGCATTGGACGAGGTCGC
ACATACAAGAGTATATAATAAATCAAGGACAGGGGGCGGGCTTTGGAACCACACTGGTCTGTCATTGGCGCTTGAA
ACTATACTACAGTTCGGACCTTTGAATATGTCAAGTGCACCGATGAGTGTCTCAACGTTGGAGAAGAGACCGAAGTG
TGTTAAAAGTGATTCGTCGAAATTGATGGTTTCGACGTCTTTAAGGTGCAAATATATTGGAGAGGTAATTGGGTTAT
CGGCCGTTTATGGTACCGAAGGTAAAGAAAAATTGATTGAATTTATAATGAAACGGGTGTGGACAGCGTGTCAGGAA

Appendix

```
CGGTCCGATTCTGAACATCGAGATGCCTTGTGGCAAGCTACGGCACTCTTGATTTCCACGACAGAGTTTCACCGGAA
TTTGTTGCATTGCATTGCGTGGTCGCAAGTCGAGCTTTTCACAGTGGAAGCAATGCGAACAGCCGTCGAATGCTGGC
AATGGCTCATAACATCACGTCCAGAACTCGAGATCCGTTTCCTCCAGGAAATGGTCTCGGCCTGGAACTGCACCGTC
CAGAAACGCATGGGTCTTTTTTCGACCAGCGAACCCATGACAAGCCCCCTAGCGGCGTACGAAGGCGCCCGACTTGA
ACCGAACCCCCCATTCGTAAAACCGCACGGAATTTGGGTCCAGTTCATTTGCGAATTGATCGACAACGTCAAATACA
GCAGTTACGAGAAAGTCGAAATGTTGGCAAGCTTGATCCATCGCTCGTTGGCAATGTGCGTGGGGGCCGACCCCCCA
TGCCAAACCCGCTCCGTTTCAGCGGTCGGTGTCCGCTTCAAGCTACTAACTTGCGGCTTATCGCTACTCCAAGGCGA
TATTTTGCCGAAAAGTCTGGCGAAAAATGTGCTCAGAGAGCGCATTTATTGCAGCTGTTTGGACTATTTCTGCAAAC
CGGTCATGTGTCCGTCGCAAACACCGACCGAATTGCGTGAGGATATCACGACACTTGTGCGATTTTGGCAAACGTTA
CACTCTGATAAGAAGTATTTGAAAGCCAGCGATGTTGGCGATTTGGACATTGGTCAGAATGCACCAATGATGGTCGA
GAATAACGAACTGACGAAACCTAACGACTTCAACCGACCAACTTCGGGTTGGATCAACACCGTACCACTGTCCAGCA
GTACCGCCACCTTAAGCAAACGCTCGGCGAAGTCAAAACGCGTCCCGATGGCCGACAACTTCGTCAAATGTTATTTG
AAAAAACGCAACCTAATTATGGACTTGTTAACGGTCGAGATTGAATTTTTGATCGTTTGGCTCAACCCGACTTCGCG
CCAGGAACAGCAAATCCCAGGCGAGGAAAACATAGCATCATGGAGGGCTAAGACCATCACCGAGAAACAATGGCAAG
ACTACACAAGGCTTGCTTGGGATATTTCGCCGGTTTTGGCCACTTATCTGCCCTCTAGATTCAAAACAAACGAAGCT
ATAATGAGCGAAATCAGGCATCAAGTGCAACAAAACCCAGTCAGTGTGTCGCACGTTCCCGAAGCGTTGCAATACCT
GGCCACCACTAACGCCATCTTGAGCGACAGCACTAAACTAGTGCACATGCTGACATGGGCCCGGGTCTCCCCTATCA
CTGCCTTGGCCTACTTCAGTCGACAATTCCCCCCGCATCCTATTACAGCACAGTATGCTGTGAAAGTGTTGAGTTCG
TATCCGGCCGATGCGGTTTTGTTTTATATTCCGCAGTTGGTGCAAGCGTTACGCCACGATACGATGGGGTATGTGAC
TGAATTTATTAAATATGTGGCGAAAAAATCGCAAATTGTGGCACATCAGTTGATTTGGAATATGAAAACTAATATGT
ATTTGGATGAGGATATGCAGCATAAGGATGTTGTCTTGTTTGATGTTTTGGATGCGTTGTGTAATAGTATTTTGGCT
GAGCTGTCGGGGCCAGCGAAACAGTTTTATGAACGAGAATTTGATTTCTTTGATAAGATTACGAGCATTTCGGGGGA
GATACGGCCCTATCCGAAGGGGCCTGAGCGCAGGAGGGCCTGTCTGGAAGCGTTGAGAAAAATTAAGGTCCAACCTG
GTTGCTACTTGCCCAGCAACCCCGAGGCAATGGTGGTCGACATCGACTACAACAGTGGCGTACCGATGCAAAGCGCC
GCCAAAGCCCCATACTTAGCCCGTTTCAAAGTGTGCCGATGCGGAATAAACGAACTCGAAAACATCGCAATGGCTGT
ATCCGTTAACGAAAACCATGAACAAAACTTCGGTCCGGAAATGTGGCAAGCGGCCATATTTAAAGTCGGCGATGACG
TCCGGCAAGACATGTTGGCACTGCAAGTGATCGGAATATTTAAGAATATTTTCCAAACGGTCGGTTTGGATTTATAT
TTGTTTCCCTATCGGGTCGTAGCCACCGCACCTGGGTGTGGGGTGATTGAATGCGTACCAAACGCCAAGTCACGTGA
CCAGTTAGGGCGGCAGACCGACATCGGACTGTATGAATATTTTTGAAAAAATACGGGGAGGAGAGTTCAAAGGAGT
TTCAAAACGCGCGCCGAAATTTCATCATATCCATGGCCGCCTACAGCGTTGTTGGCTTTCTTTTACAAATCAAAGAC
CGTCACAATGGCAATATTATGTTGGACACGGACGGTCATATAATCCACATTGATTTTGGTTTCATGTTCGAGTCATC
GCCAGGCGGTAACTTGGGTTTTGAGCCCGATATTAAGCTCACCGATGAGATGGTTATGATAATGGGGGGCAAAATGG
AAGCAGCCCCCTTTAAATGGTTTTCGGACCTGTGTGTCCAGGCTTATTTGGCCGTCAGGCCTTATCAGGAGTCCATC
ATTTCGCTGGTCTCGCTCATGTTGGACACGGGGCTGCCCTGCTTTCGGGGGCAGACGATCAAGTTGTTAAGGGGCG
CTTCCATCCGGGGGCCACCGATAAGGAAGCGGCCACTTCATGCTGCAAATCATACGTAACAGTTTCCTGAATTTCA
GGACAAGGACCTATGACATGATCCAATATTATCAGAATCAGATTCCTTATTAA
```

Appendix

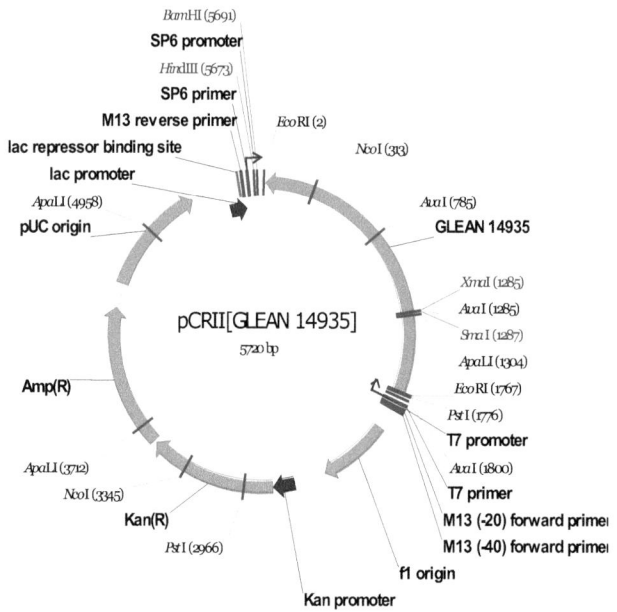

>pCRII[GLEAN_14935]

GAATTCGGCTTTGTCCAACATGAGCGAGACCAGCGAAATGATGGACTCCTGATAAGGCCTGACGGCCAAATAAGCCT
GGACACACAGGTCCGAAAACCATTTAAAGGGGGCTGCTTCCATTTTGCCCCCCATTATCATAACCATCTCATCGGTG
AGCTTAATATCGGGCTCAAAACCCAAGTTACCGCCTGGCGATGACTCGAACATGAAACCAAAATCAATGTGGATTAT
ATGACCGTCCGTGTCCAACATAATATTGCCATTGTGACGGTCTTTGATTTGTAAAAGAAAGCCAACAACGCTGTAGG
CGGCCATGGATATGATGAAATTTCGGCGCGCGTTTTGAAACTCCTTTGAACTCTCCTCCCCGTATTTTTTCAAAAAA
TATTCATACAGTCCGATGTCGGTCTGCCGCCCTAACTGGTCACGTGACTTGGCGTTTGGTACGCATTCAATCACCCC
ACACCCAGGTGCGGTGGCTACGACCCGATAGGGAAACAAATATAAATCCAAACCGACCGTTTGGAAAATATTCTTAA
ATATTCCGATCACTTGCAGTGCCAACATGTCTTGCCGGACGTCATCGCCGACTTTAAATATGGCCGCTTGCCACATT
TCCGGACCGAAGTTTTGTTCATGGTTTTCGTTAACGGATACAGCCATTGCGATGTTTCGAGTTCGTTTATTCCGCA
TCGGCACACTTTGAAACGGGCTAAGTATGGGGCTTTGGCGGCGCTTTGCATCGGTACGCCACTGTTGTAGTCGATGT
CGACCACCATTGCCTCGGGGTTGCTGGGCAAGTAGCAACCAGGTTGGACCTTAATTTTTCTCAACGCTTCCAGACAG
GCCCTCCTGCGCTCAGGCCCCTTCGGATAGGGCCGTATCTCCCCCGAAATGCTCGTAATCTTATCAAAGAAATCAAA
TTCTCGTTCATAAAACTGTTTCGCTGGCCCCGACAGCTCAGCCAAATACTATTACACAACGCATCCAAAACATCAA
ACAAGACAACATCCTTATGCTGCATATCCTCATCCAAATACATATTAGTTTTCATATTCCAAATCAACTGATGTGCC
ACAATTTGCGATTTTTTCGCCACATATTTAATAAATTCAGTCACATACCCCATCGTATCGTGGCGTAACGCTTGCAC
CAACTGCGGAATATAAAACAAAACCGCATCGGCCGGATACGAACTCAACACTTTCACAGCATACTGTGCTGTAATAG
GATGCGGGGGGAATTGTCGACTGAAGTAGGCCAAGGCAGTGATAGGGGAGACCCGGGCCCATGTCAGCATGTGCACT
AGTTTAGTGCTGTCGCTCAAGATGGCGTTAGTGGTGGCCAGGTATTGCAACGCTTCGGGAACGTGCGACACACTGAC
TGGGTTTTGTTGCACTTGATGCCTGATTTCGCTCATTATAGCTTCGTTTGTTTTGAATCTAGAGGGCAGATAAGTGG
CCAAAACCGGCGAAATATCCCAAGCAAGCCTTGTGTAGTCTTGCCATTGTTTCTCGGTGATGGTCTTAGCCCTCCAT
GATGCTATGTTTTCCTCGCCTGGGATTTGCTGTTCCTGGCGCGAAGTCGGGTTGAGCCAAACGATCAAAAATTCAAT
CTCGACCGTTAACAAGTCCATAATTAGGTTGCGTTTTTTCAAATAACATTTGACGAAGTTGTCGGCCATCGGGACGC
GTTTTGACTTCGCCGAGCGTTTGCTTAAGGTGGCGGTACTGCTGGACAGTGGTACGGTGTTGATCCAAGCCGAATTC
TGCAGATATCCATCACACTGGCGGCCGCTCGAGCATGCATCTAGAGGGCCCAATTCGCCCTATAGTGAGTCGTATTA
CAATTCACTGGCCGTCGTTTTACAACGTCGTGACTGGGAAAACCCTGGCGTTACCCAACTTAATCGCCTTGCAGCAC
ATCCCCCTTTCGCCAGCTGGCGTAATAGCGAAGAGGCCCGCACCGATCGCCCTTCCCAACAGTTGCGCAGCCTGAAT
GGCGAATGGACGCGCCCTGTAGCGGCGCATTAAGCGCGGCGGGTGTGGTGGTTACGCGCAGCGTGACCGCTACACTT
GCCAGCGCCCTAGCGCCCGCTCCTTTCGCTTTCTTCCCTTCCTTTCTCGCCACGTTCGCCGGCTTTCCCCGTCAAGC
TCTAAATCGGGGGCTCCCTTTAGGGTTCCGATTTAGTGCTTTACGGCACCTCGACCCCAAAAAACTTGATTAGGGTG

134

Appendix

```
ATGGTTCACGTAGTGGGCCATCGCCCTGATAGACGGTTTTTCGCCCTTTGACGTTGGAGTCCACGTTCTTTAATAGT
GGACTCTTGTTCCAAACTGGAACAACACTCAACCCTATCTCGGTCTATTCTTTTGATTTATAAGGGATTTTGCCGAT
TTCGGCCTATTGGTTAAAAAATGAGCTGATTTAACAAAAATTTAACGCGAATTTTAACAAAATTCAGGGCGCAAGGG
CTGCTAAAGGAAGCGGAACACGTAGAAAGCCAGTCCGCAGAAACGGTGCTGACCCCGGATGAATGTCAGCTACTGGG
CTATCTGGACAAGGGAAAACGCAAGCGCAAAGAGAAAGCAGGTAGCTTGCAGTGGGCTTACATGGCGATAGCTAGAC
TGGGCGGTTTTATGGACAGCAAGCGAACCGGAATTGCCAGCTGGGGCGCCCTCTGGTAAGGTTGGGAAGCCCTGCAA
AGTAAACTGGATGGCTTTCTTGCCGCCAAGGATCTGATGGCGCAGGGGATCAAGATCTGATCAAGAGACAGGATGAG
GATCGTTTCGCATGATTGAACAAGATGGATTGCACGCAGGTTCTCCGGCCGCTTGGGTGGAGAGGCTATTCGGCTAT
GACTGGGCACAACAGACAATCGGCTGCTCTGATGCCGCCGTGTTCCGGCTGTCAGCGCAGGGGCGCCCGGTTCTTTT
TGTCAAGACCGACCTGTCCGGTGCCCTGAATGAACTGCAGGACGAGGCAGCGCGGCTATCGTGGCTGGCCACGACGG
GCGTTCCTTGCGCAGCTGTGCTCGACGTTGTCACTGAAGCGGGAAGGGACTGGCTGCTATTGGGCGAAGTGCCGGGG
CAGGATCTCCTGTCATCCCACCTTGCTCCTGCCGAGAAAGTATCCATCATGGCTGATGCAATGCGGCGGCTGCATAC
GCTTGATCCGGCTACCTGCCCATTCGACCACCAAGCGAAACATCGCATCGAGCGAGCACGTACTCGGATGGAAGCCG
GTCTTGTCGATCAGGATGATCTGGACGAAGAGCATCAGGGGCTCGCGCCAGCCGAACTGTTCGCCAGGCTCAAGGCG
CGCATGCCCGACGGCGAGGATCTCGTCGTGACCCATGGCGATGCCTGCTTGCCGAATATCATGGTGGAAAATGGCCG
CTTTTCTGGATTCATCGACTGTGGCCGGCTGGGTGTGGCGGACCGCTATCAGGACATAGCGTTGGCTACCCGTGATA
TTGCTGAAGAGCTTGGCGGCGAATGGGCTGACCGCTTCCTCGTGCTTTACGGTATCGCCGCTCCCGATTCGCAGCGC
ATCGCCTTCTATCGCCTTCTTGACGAGTTCTTCTGAATTGAAAAAGGAAGAGTATGAGTATTCAACATTTCCGTGTC
GCCCTTATTCCCTTTTTTGCGGCATTTTGCCTTCCTGTTTTTGCTCACCCAGAAACGCTGGTGAAAGTAAAAGATGC
TGAAGATCAGTTGGGTGCACGAGTGGGTTACATCGAACTGGATCTCAACAGCGGTAAGATCTTGAGAGTTTTCGCC
CCGAAGAACGTTTTCCAATGATGAGCACTTTTAAAGTTCTGCTATGTGGCGCGGTATTATCCCGTATTGACGCCGGG
CAAGAGCAACTCGGTCGCCGCATACACTATTCTCAGAATGACTTGGTTGAGTACTCACCAGTCACAGAAAAGCATCT
TACGGATGGCATGACAGTAAGAGAATTATGCAGTGCTGCCATAACCATGAGTGATAACACTGCGGCCAACTTACTTC
TGACAACGATCGGAGGACCGAAGGAGCTAACCGCTTTTTTGCACAACATGGGGGATCATGTAACTCGCCTTGATCGT
TGGGAACCGGAGCTGAATGAAGCCATACCAAACGACGAGCGTGACACCACGATGCCTGTAGCAATGGCAACAACGTT
GCGCAAACTATTAACTGGCGAACTACTTACTCTAGCTTCCCGGCAACAATTAATAGACTGGATGGAGGCGGATAAAG
TTGCAGGACCACTTCTGCGCTCGGCCCTTCCGGCTGGCTGGTTTATTGCTGATAAATCTGGAGCCGGTGAGCGTGGG
TCTCGCGGTATCATTGCAGCACTGGGGCCAGATGGTAAGCCCTCCCGTATCGTAGTTATCTACACGACGGGGAGTCA
GGCAACTATGGATGAACGAAATAGACAGATCGCTGAGATAGGTGCCTCACTGATTAAGCATTGGTAACTGTCAGACC
AAGTTTACTCATATATACTTTAGATTGATTTAAAACTTCATTTTTAATTTAAAAGGATCTAGGTGAAGATCCTTTTT
GATAATCTCATGACCAAAATCCCTTAACGTGAGTTTTCGTTCCACTGAGCGTCAGACCCCGTAGAAAAGATCAAAGG
ATCTTCTTGAGATCCTTTTTTTCTGCGCGTAATCTGCTGCTTGCAAACAAAAAAACCACCGCTACCAGCGGTGGTTT
GTTTGCCGGATCAAGAGCTACCAACTCTTTTTCCGAAGGTAACTGGCTTCAGCAGAGCGCAGATACCAAATACTGTT
CTTCTAGTGTAGCCGTAGTTAGGCCACCACTTCAAGAACTCTGTAGCACCGCCTACATACCTCGCTCTGCTAATCCT
GTTACCAGTGGCTGCTGCCAGTGGCGATAAGTCGTGTCTTACCGGGTTGGACTCAAGACGATAGTTACCGGATAAGG
CGCAGCGGTCGGGCTGAACGGGGGGTTCGTGCACACAGCCCAGCTTGGAGCGAACGACCTACACCGAACTGAGATAC
CTACAGCGTGAGCTATGAGAAAGCGCCACGCTTCCCGAAGGGAGAAAGGCGGACAGGTATCCGGTAAGCGGCAGGGT
CGGAACAGGAGAGCGCACGAGGGAGCTTCCAGGGGGAAACGCCTGGTATCTTTATAGTCCTGTCGGGTTTCGCCACC
TCTGACTTGAGCGTCGATTTTTGTGATGCTCGTCAGGGGGCGGAGCCTATGGAAAAACGCCAGCAACGCGGCCTTT
TTACGGTTCCTGGCCTTTTGCTGGCCTTTTGCTCACATGTTCTTTCCTGCGTTATCCCCTGATTCTGTGGATAACCG
TATTACCGCCTTTGAGTGAGCTGATACCGCTCGCCGCAGCCGAACGACCGAGCGCAGCGAGTCAGTGAGCGAGGAAG
CGGAAGAGCGCCCAATACGCAAACCGCCTCTCCCCGCGCGTTGGCCGATTCATTAATGCAGCTGGCACGACAGGTTT
CCCGACTGGAAAGCGGGCAGTGAGCGCAACGCAATTAATGTGAGTTAGCTCACTCATTAGGCACCCCAGGCTTTACA
CTTTATGCTTCCGGCTCGTATGTTGTGTGGAATTGTGAGCGGATAACAATTTCACACAGGAAACAGCTATGACCATG
ATTACGCCAAGCTATTTAGGTGACACTATAGAATACTCAAGCTATGCATCAAGCTTGGTACCGAGCTCGGATCCACT
AGTAACGGCCGCCAGTGTGCTGT
```

Appendix

G10215:

*piggy*Bac left arm flanking sequence:

TTAAAAGAGCTTTTACTTAAAAGAAGCACTAAAAAGTAATAAATGATGTTTACTTATCACAGAAGAATATTTTTAG
TAACAGATTAGATTCTAAAAATTATAAAAGCTTTAGGAGTGTCATAGTCGTAACTCAGAACACTACGGCATTACAGA
ATTTTTGCTTTGTTCAATTGACGTCTTCCAAAGCTTTTTATAGGAAATATAGAACAGACAGCAAAGCAAGTCAAATA
AAGCTTTTAAAAAGGAGTCATATTACAGCTGTATACGCATTATCGGAAACCAAACAATCATAATATGTATTTTCATA
ATAGAAAACTGAAAATAATAACAATTGAAGTCCAAAGGGTCCAGCTAGTTAACTTGGAAAAATTAATTTGTTGGAAA
AAGTATGTTTAAAATTCTATAACTCATTCCACTACCTTATCTAAACATTCGTTAGATC

GLEAN_00032 DNA sequence (the cloned sequence is marked in yellow):

ATGATTGTACAGGAACCAGTGCAGGCTGCCATTTGGCATTGTTTGAATCATTATGACTACACTGATGCTGTATTCTT
GTCAGAGCGATTATATGCTGAAGTCAAATCTGATGATAGTCTGTACCTCTTGGCCACAGCTTATTACAGAAGCGGCC
AAAAAGACCACGCATACCACATTTTGAAGGAAAGGACAGATGCTTCAACACAGTGCAGATATTTGCTTGGGATTTGT
GCTTATGATCTTGAAAAATATGCTGAGGCCGAGGCAGCTTTGTTACACTCCAATAAGAGCAGCAATGACTCCGAGAA
TTTCGACGACATCACATCAGAATACGGTGATCAAGCACCATTCGCTTTATCTTTATTAGGTAACATAGCTGCCAAAA
CCGAAAGAAAACCTCGTGCCATAGATGCATGGAAACGAGCCTTAAAACTAAATCCATTTCAATGGTCAAGCTTCGAA
AATTTATGCAAAATAGGAGATAAACCAAACCCCCAAAATATATTCCAAATAACAGGTGTTGAAAATTTATCAATGTG
CCAGGGAAGTAATATTAACAATATCGAAAGTGTTGTTATAACAAACAACAATCCAAATCAAGATAATCAAGAGACAT
ACGCCACAACTCCTCAACAGATTCTTGTGAATTTTAGCCCCAATGTCGTAACTTCGAATTCGAAATTGTGTACACCT
GATGATAATCCAATTGTTACACAAAATGTGTGGCTGTCGGGCTTCGCCCCCTTCACGTCGAGGACCAAGTTGCTGAG
ATTTCGGACGGATTCCTCCAGTAATCCGAGTCCTCAGCCGAGCTTTGGCCACTTCTTGGACACCTCCCCGACGGTT
ACTTCAGCACACCGAACCTCCTATCCCAACCAAAACTGACCGAATCCAACAACCACCACCAATCGCTAGCAAAGCGC
GTGAAAGCGCAAGTCGATCAATTCATCAATCGCAAAGAACCGATCCTCCAAAACTGCAAACCCCTATTCAGCCAATC
GACAAACGTCACCAAAACCCCAACAATCTCCCTTGCCGTACAGCCATGTCAAAACGTCAGACGCAGCTCACGCCTCT
TCACCAACAGCTACTCAGTCAAAGAAAACAACAAATCGCCCAATCGAACCAAATTCGCCACACCCAAGTCGCCCTCG
CGGAAAACCAAACAACGAATGGCGAAATGCAATTTAAACAAAACCACAGCCTACACCGAACTCAACACACGCAACCG
CATCGAGAAGGAAAAAACCGAAACGGTGACTTCGACCGACGCAAAAATCGAGAAAAGTGCAATAAGTAATAGTAGTA
GTGTCGAGAGTTGCGTCCAGCAGGCGATTCTAATGCAGAAGCAAAGTGCCGAAGGGCTGATGGTCCTCTTGCGGAGT
TTAGGTCAGGCGTACTTGCATTTGTCGAATTTTAATTGCAAGGCGGCGATTGAAGAATTGAACGTTCTGCCGCCGAA
TCAGTTCCAAACGGCTTGGATTTATTGTCTTCTGGGGTTAGCCTACTTCGAATTGACCGACTACGAGTCCAGCATAA
AGTACTTCAGTAAAGTGCACAACCTTGAACCGTACCGGATACAGTTCATGGACGTCTACTCCACAGCGTTGTGGCAC
TTGCAGAAGGAGGTGGCGTTGTCAGCGCTGGCCCAGGACCTCATCTCTTTGAACAAAAACTCGCCGGTGACGTGGTG
CGTCAGCGGAAACTGCTTCTCGTTGCATAAGGAGCACGACACCGCGATTAAGTTTTTCCAAAGGGCTGTGCAAGTCG
ATCCGAGGTTTCCCTACGCGTATACGCTGCTAGGGCATGAGTACATCACGACGGAGGAGCTTGACAAGGCGATGAGT
TGCTTCCGCAATGCCATAAGACTGGACCCAAGACATTACAATGCCTGGTTTGGTATCGGAACCATTTACTCAAAACA
GGAGAGGTACCATCTGGCGGAAATAAACTACTCGCGAGCGCTTGAGATCAACCCCCAGAGCTCGGTGATTCTTTGCC
ACATTGGTATCGTCCAACACGCGTTGAAGCAGACGGAGAAGGCTCTGAAGACGTTTAACGTGGCCATTGCGAATAAT
CCGAAGAGTCCGCTGTGTAAGTTCCACCGGGGGTCGATTTATTTCGCCCTCGGGAGGCACGCAGAGGCCTTAAAGGA
GCTGGAGGAGTTGAAGGAGATCGTTCCCAAAGAGTCGCTAGTTTACTATTTAATAGGCAAAGTTCACAAGAAGTTAG
GAAACACGGATTTAGCGCTAATGCATTTCAGTTGGGCGACGGATTTGGACCCGAAAGGGGCGAGTAGTCAGATCAAG
GAGGCGTTTGATCCGTCGATCGGTCGCACTACGTCCAGTGAGTCGCCCACTTCGCCAGCGGCGGAGGAGTTTCTGCC
GGATAGGAGGCCGGGCTTGCAGCCGGAAGCGTTGAACTTCGGGAATCTTCCAGACGATAGCGAGGATAGTTTCTAA

Appendix

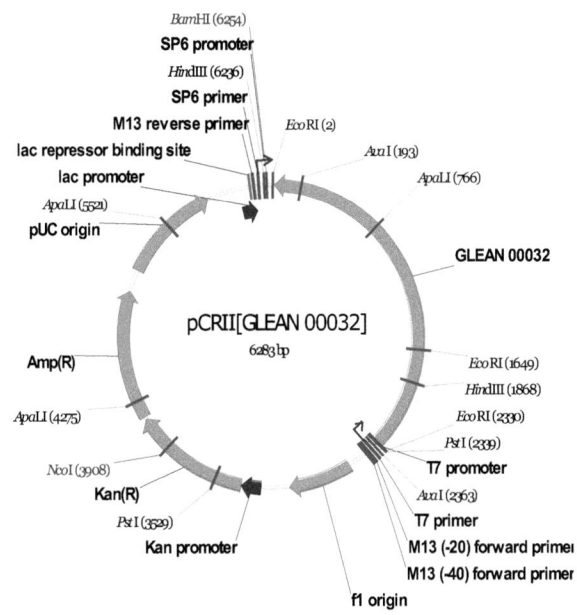

>pCRII[GLEAN_00032]

```
GAATTCGGCTTCTCCTTGATCTGACTACTCGCCCCTTTCGGGTCCAAATCCGTCGCCCAACTGAAATGCATTAGCGC
TAAATCCGTGTTTCCTAACTTCTTGTGAACTTTGCCTATTAAATAGTAAACTAGCGACTCTTTGGGAACGATCTCCT
TCAACTCCTCCAGCTCCTTTAAGGCCTCTGCGTGCCTCCCGAGGGCGAAATAAATCGACCCCCGGTGGAACTTACAC
AGCGGACTCTTCGGATTATTCGCAATGGCCACGTTAAACGTCTTCAGAGCCTTCTCCGTCTGCTTCAACGCGTGTTC
GACGATACCAATGTGGCAAAGAATCACCGAGCTCTGGGGGTTGATCTCAAGCGCTCGCGAGTAGTTTATTTCCGCCA
GATGGTACCTCTCCTGTTTTGAGTAAATGGTTCCGATACCAAACCAGGCATTGTAATGTCTTGGGTCCAGTCTTATG
GCATTGCGGAAGCAACTCATCGCCTTGTCAAGCTCCTCCGTCGTGATGTACTCATGCCCTAGCAGCGTATACGCGTA
GGGAAACCTCGGATCGACTTGCACAGCCCTTTGGAAAAACTTAATCGCGGTGTCGTGCTCCTTATGCAACGAGAAGC
AGTTTCCGCTGACGCACCACGTCACCGGCGAGTTTTTGTTCAAAGAGATGAGGTCCTGGGCCAGCGCTGACAACGCC
ACCTCCTTCTGCAAGTGCCACAACGCTGTGGAGTAGACGTCCATGAACTGTATCCGGTACGGTTCAAGGTTGTGCAC
TTTACTGAAGTACTTTATGCTGGACTCGTAGTCGGTCAATTCGAAGTAGGCTAACCCCAGAAGACAATAAATCCAAG
CCGTTTGGAACTGATTCGGCGGCAGAACGTTCAATTCTTCAATCGCCGCCTTGCAATTAAAATTCGACAAATGCAAG
TACGCCTGACCTAAACTCCGCAAGAGGACCATCAGCCCTTCGGCACTTTGCTTCTGCATTAGAATGCCTGCTGGAC
GCAACTCTCGACACTACTACTATTACTTATTGCACTTTTCTCGATTTTTGCGTCGGTCGAAGTCACCGTTTCGGTTT
TTTCCTTCTCGATGCGGTTGCGTGTGTTGAGTTCGGTGTAGGCTGTGGTTTTGTTTAAATTGCATTTCGCCATTCGT
TGTTTGGTTTTCCGCGAGGGCGACTTGGGTGTGGCGAATTTGGTTCGATTGGGCGATTTGTTGTTTTCTTTGACTGA
GTAGCTGTTGGTGAAGAGGCGTGAGCTGCGTCTGACGTTTTGACATGGCTGTACGGCAAGGGAGATTGTTGGGGTTT
TGGTGACGTTTGTCGATTGGCTGAATAGGGGTTTGCAGTTTTGGAGGATCGGTTCTTTGCGATTGATGAATTGATCG
ACTTGCGCTTTCACGCGCTTTGCTAGCGATTGGTGGTGGTTGTTGGATTCGGTCAGTTTTGGTTGGGATAGGAGGTT
CGGTGTGCTGAAGTAACCGTCGGGGGAGGTGTCCAAGAAGTGGCCAAAGCTCGGCTGAGGACTCGGATTACTGGAGG
AATCCGTCCGAAATCTCAGCAACTTGGTCCTCGACGTGAAGGGGGCGAAGCCCGACAGCCACACATTTTGTGTAACA
ATTGGATTATCATCAGGTGTACACAATTTCGAATTCGAAGTTACGACATTGGGCTAAAATTCACAAGAATCTGTTG
AGGAGTTGTGGCGTATGTCTCTTGATTATCTTGATTTGGATTGTTGTTTGTTATAACAACACTTTCGATATTGTTAA
TATTACTTCCCTGGCACATTGATAAATTTTCAACACCTGTTATTTGGAATATATTTTGGGGGTTTGGTTTATCTCCT
ATTTTGCATAAATTTTCGAAGCTTGACCATTGAAATGGATTTAGTTTTAAGGCTCGTTTCCATGCATCTATGGCACG
AGGTTTTCTTTCGGTTTTGGCAGCTATGTTACCTAATAAAGATAAAGCGAATGGTCTTGATCACCGTATTCTGATG
TGATGTCGTCGAAATTCTCGGAGTCATTGCTGCTCTTATTGGAGTGTAACAAAGCTGCCTCGGCCTCAGCATATTTT
TCAAGATCATAAGCACAAATCCCAAGCAAATATCTGCACTGTGTTGAAGCATCTGTCCTTTCCTTCAAAATGTGGTA
TGCGTGGTCTTTTTGGCCGCTTCTGTAATAAGCTGTGGCCAAGAGGTACAGACTATCATCAGATTTGACTTCAGCAT
```

137

Appendix

```
ATAATCGCTCTGACAAGAATACAGCATCAGTGTAGTCATAATGATTCAAACAATGCCAAATGGCAGCCTGCACTGGT
TCCTGTACAATCAAAGCCGAATTCTGCAGATATCCATCACACTGGCGGCCGCTCGAGCATGCATCTAGAGGGCCCAA
TTCGCCCTATAGTGAGTCGTATTACAATTCACTGGCCGTCGTTTTACAACGTCGTGACTGGGAAAACCCTGGCGTTA
CCCAACTTAATCGCCTTGCAGCACATCCCCCTTTCGCCAGCTGGCGTAATAGCGAAGAGGCCCGCACCGATCGCCCT
TCCCAACAGTTGCGCAGCCTGAATGGCGAATGGACGCGCCCTGTAGCGGCGCATTAAGCGCGGCGGGTGTGGTGGTT
ACGCGCAGCGTGACCGCTACACTTGCCAGCGCCCTAGCGCCCGCTCCTTTCGCTTTCTTCCCTTCCTTTCTCGCCAC
GTTCGCCGGCTTTCCCCGTCAAGCTCTAAATCGGGGGCTCCCTTTAGGGTTCCGATTTAGTGCTTTACGGCACCTCG
ACCCCAAAAAACTTGATTAGGGTGATGGTTCACGTAGTGGGCCATCGCCCTGATAGACGGTTTTTCGCCCTTTGACG
TTGGAGTCCACGTTCTTTAATAGTGGACTCTTGTTCCAAACTGGAACAACACTCAACCCTATCTCGGTCTATTCTTT
TGATTTATAAGGGATTTTGCCGATTTCGGCCTATTGGTTAAAAAATGAGCTGATTTAACAAAAATTTAACGCGAATT
TTAACAAAATTCAGGGCGCAAGGGCTGCTAAAGGAAGCGGAACACGTAGAAAGCCAGTCCGCAGAAACGGTGCTGAC
CCCGGATGAATGTCAGCTACTGGGCTATCTGGACAAGGGAAAACGCAAGCGCAAAGAGAAAGCAGGTAGCTTGCAGT
GGGCTTACATGGCGATAGCTAGACTGGGCGGTTTTATGGACAGCAAGCGAACCGGAATTGCCAGCTGGGGCGCCCTC
TGGTAAGGTTGGGAAGCCCTGCAAAGTAAACTGGATGGCTTTCTTGCCGCCAAGGATCTGATGGCGCAGGGGATCAA
GATCTGATCAAGAGACAGGATGAGGATCGTTTCGCATGATTGAACAAGATGGATTGCACGCAGGTTCTCCGGCCGCT
TGGGTGGAGAGGCTATTCGGCTATGACTGGGCACAACAGACAATCGGCTGCTCTGATGCCGCCGTGTTCCGGCTGTC
AGCGCAGGGGCGCCCGGTTCTTTTTGTCAAGACCGACCTGTCCGGTGCCCTGAATGAACTGCAGGACGAGGCAGCGC
GGCTATCGTGGCTGGCCACGACGGGCGTTCCTTGCGCAGCTGTGCTCGACGTTGTCACTGAAGCGGGAAGGGACTGG
CTGCTATTGGGCGAAGTGCCGGGGCAGGATCTCCTGTCATCCCACCTTGCTCCTGCCGAGAAAGTATCCATCATGGC
TGATGCAATGCGGCGGCTGCATACGCTTGATCCGGCTACCTGCCCATTCGACCACCAAGCGAAACATCGCATCGAGC
GAGCACGTACTCGGATGGAAGCCGGTCTTGTCGATCAGGATGATCTGGACGAAGAGCATCAGGGGCTCGCGCCAGCC
GAACTGTTCGCCAGGCTCAAGGCGCGCATGCCCGACGGCGAGGATCTCGTCGTGACCCATGGCGATGCCTGCTTGCC
GAATATCATGGTGGAAAATGGCCGCTTTTCTGGATTCATCGACTGTGGCCGGCTGGGTGTGGCGGACCGCTATCAGG
ACATAGCGTTGGCTACCCGTGATATTGCTGAAGAGCTTGGCGGCGAATGGGCTGACCGCTTCCTCGTGCTTTACGGT
ATCGCCGCTCCCGATTCGCAGCGCATCGCCTTCTATCGCCTTCTTGACGAGTTCTTCTGAATTGAAAAGGAAGAGT
ATGAGTATTCAACATTTCCGTGTCGCCCTTATTCCCTTTTTTGCGGCATTTTGCCTTCCTGTTTTTGCTCACCCAGA
AACGCTGGTGAAAGTAAAAGATGCTGAAGATCAGTTGGGTGCACGAGTGGGTTACATCGAACTGGATCTCAACAGCG
GTAAGATCCTTGAGAGTTTTCGCCCCGAAGAACGTTTTCCAATGATGAGCACTTTTAAAGTTCTGCTATGTGGCGCG
GTATTATCCCGTATTGACGCCGGGCAAGAGCAACTCGGTCGCCGCATACACTATTCTCAGAATGACTTGGTTGAGTA
CTCACCAGTCACAGAAAAGCATCTTACGGATGGCATGACAGTAAGAGAATTATGCAGTGCTGCCATAACCATGAGTG
ATAACACTGCGGCCAACTTACTTCTGACAACGATCGGAGGACCGAAGGAGCTAACCGCTTTTTTGCACAACATGGGG
GATCATGTAACTCGCCTTGATCGTTGGGAACCGGAGCTGAATGAAGCCATACCAAACGACGAGCGTGACACCACGAT
GCCTGTAGCAATGGCAACAACGTTGCGCAAACTATTAACTGGCGAACTACTTACTCTAGCTTCCCGGCAACAATTAA
TAGACTGGATGGAGGCGGATAAAGTTGCAGGACCACTTCTGCGCTCGGCCCTTCCGGCTGGCTGGTTTATTGCTGAT
AAATCTGGAGCCGGTGAGCGTGGGTCTCGCGGTATCATTGCAGCACTGGGGCCAGATGGTAAGCCCTCCCGTATCGT
AGTTATCTACACGACGGGGAGTCAGGCAACTATGGATGAACGAAATAGACAGATCGCTGAGATAGGTGCCTCACTGA
TTAAGCATTGGTAACTGTCAGACCAAGTTTACTCATATATACTTTAGATTGATTTAAAACTTCATTTTTAATTTAAA
AGGATCTAGGTGAAGATCCTTTTTGATAATCTCATGACCAAAATCCCTTAACGTGAGTTTTCGTTCCACTGAGCGTC
AGACCCCGTAGAAAAGATCAAAGGATCTTCTTGAGATCCTTTTTTTTCTGCGCGTAATCTGCTGCTTGCAAACAAAAA
AACCACCGCTACCAGCGGTGGTTTGTTTGCCGGATCAAGAGCTACCAACTCTTTTTCCGAAGGTAACTGGCTTCAGC
AGAGCGCAGATACCAAATACTGTTCTTCTAGTGTAGCCGTAGTTAGGCCACCACTTCAAGAACTCTGTAGCACCGCC
TACATACCTCGCTCTGCTAATCCTGTTACCAGTGGCTGCTGCCAGTGGCGATAAGTCGTGTCTTACCGGGTTGGACT
CAAGACGATAGTTACCGGATAAGGCGCAGCGGTCGGGCTGAACGGGGGGTTCGTGCACACAGCCCAGCTTGGAGCGA
ACGACCTACACCGAACTGAGATACCTACACAGCGTGAGCTATGAGAAAGCGCCACGCTTCCCGAAGGGAGAAAGGCGGA
CAGGTATCCGGTAAGCGGCAGGGTCGGAACAGGAGAGCGCACGAGGGAGCTTCCAGGGGGAAACGCCTGGTATCTTT
ATAGTCCTGTCGGGTTTCGCCACCTCTGACTTGAGCGTCGATTTTTGTGATGCTCGTCAGGGGGGCGGAGCCTATGG
AAAAACGCCAGCAACGCGGCCTTTTTACGGTTCCTGGCCTTTTGCTGGCCTTTTGCTCACATGTTCTTTCCTGCGTT
ATCCCCTGATTCTGTGGATAACCGTATTACCGCCTTTGAGTGAGCTGATACCGCTCGCCGCAGCCGAACGACCGAGC
GCAGCGAGTCAGTGAGCGAGGAAGCGGAAGAGCGCCCAATACGCAAACCGCCTCTCCCCGCGCGTTGGCCGATTCAT
TAATGCAGCTGGCACGACAGGTTTCCCGACTGGAAAGCGGGCAGTGAGCGCAACGCAATTAATGTGAGTTAGCTCAC
TCATTAGGCACCCCAGGCTTTACACTTTATGCTTCCGGCTCGTATGTTGTGTGGAATTGTGAGCGGATAACAATTTC
ACACAGGAAACAGCTATGACCATGATTACGCCAAGCTATTTAGGTGACACTATAGAATACTCAAGCTATGCATCAAG
CTTGGTACCGAGCTCGGATCCACTAGTAACGGCCGCCAGTGTGCTG
```

Appendix

KT1269:

*piggy*Bac left arm flanking sequence:

```
TTAAAATACGCCAAAAAATTGTCGAATATTGTGCCTAAAACGGTGCTATTTTCACTTTTATAGGGCAGGAAACCAAG
AATTTCATTAAAAATAAGTAGAAAATTTAAGTTGATCTAGTTTTACCTTTTTCAAACGTTTTACCTCTGCCGAAAAA
TCACTAATTTGTGCCTTTTCGACAGTTTTAACGCATTCCTGGTTAATATTTCAGAATTATTTTGCATTTTTAGAAAA
CTTGTGCTAGAACTGGGTTTGCTATTTCATTCAGATAAATTGACTGAAAAAAGTTTAGGAGATTACTGAATTTTAAA
ATGTGTGTAATATCACCGAAAAATGCATAATTTTAAACTTTGCGAACATTGTTTCTTACTAAATGATCCGGCTCGGA
GGACCATACGTTGGACCACAGATACAGTTCTACATTTCTAGGAGTATCGTGATTGAACGGTGATGAGCAAAAACTTG
GGGTTAAGTTCAAAATTGATTTTATTACAAAATTTTCCTACAAATGAAAACAAAAATATAACGCACGAATGCAGACA
ATGTAC
```

GLEAN_13098:

```
ATGGGGGGCGACACCTCCCCCGAGCAGCAGTACTCGCTGCGTTGGAATGACTTCCACTCGTCGATCCTCTCCTCGTT
TCGACACTTGAGGGACGAAGAAGATTTCGTCGATGTGACTTTGGCCTGTGATGGATGTTCGTTTACGGCACACAAAG
TTGTGCTGTCGGCGTGCAGTCCCTACTTTAGGAGGTTGCTCAAAGCCAATCCCTGCCAGCATCCCATCGTCATCCTC
AGGGACGTACAGCAAAAAGACATGGAAAGTCTGTTACGATTTATGTACAACGGAGAAGTTCACATAGGACAAGAACA
ATTAACGGATTTTTTAAAGACTGCACAAATGTTACAAGTGAGGGACTAGCCGACGTACCGGCGGGTACTGCAGGAC
AAAGACTGACTGCTCCAGAGCAAAAGATCTCACCTAATTCTTCATCTTTACCGTGGGCGACGGACCGATCGGATGCT
CTCCGCGAGGGCGCTTTATCGCCGCCTCCGGCCAAAAGGTCCAGGAGTTCGGAGCGGGGGACGTACACCCCCGACAG
GGGCCGAAGTCCCTCCAGAGGCAACGGCGACATGCAGGAATCCCTCCTCGGACAGGCCTTGGAAGGCGGCCCCACGA
TACTCACAAAGGAGAAGGGAAATTCGGATTCGTCGCTGCAAGCGCAATCGACGGGCGAGGACAGCAACAGCAGCGAC
ACGGCGATGAGCGACCATGGAGACCCTGTAACCCCAAAGACTGAACCTTCCGACTACCCCATGCTAGACGACCACCA
TCCTTTCAATACCAACGGCGGGCTAATTGATCAGACGCGGACTCCGACATTTCCAGGCGCGCTGTTAGGTCTACAAG
GATTAGGAGGACTTATGCCAGGACCGTCAGGGATCCATAATAGTACGGACAATTTCGGGGCGGCACAAGCAAATGCG
GCTAATTGTTCCTGGCAGGAGAAGCCGAAATGTCAATACTGCGGCCGTCTTTACTCAAACCCCAGTAACTTAAGACA
GCACATCATAAACGTGCACGTGCAGACGTCGCAAGAGAACTGGATCGTGTGCAACTTCTGCGGCAAAAAGTGCAAGA
CGAAGCACTACTTGATCAACCACCAGCTCCAAGCGCACGGCATCAGACAGCGCCAGCACGCGCCCACTTATTAG
```

GLEAN_13099

```
ATGTCGCGACAACATAGAGGAATAAGTACAAAAGACCTGCCTGTTCTACCAATGCCCGCACCGTTCGACCCGGAACT
TGCCAGCAGGTTGTTGGCAAAAGCCGGCGTGAAAATAAGTCCGGCCGAGCTGCGAGCGCGCGCGTCTCCGACGGGTC
CGCGTCGCTCCGACGTGAAGTTGGACGCGAAAAGCGCGTACTCTGGCGCGGCCTCGGAGGCCGGCAGCTCAATATGC
GGCGACAACGACCCCGAAGACCTGACAATCTCGCACCAAAGGTATGGAGGGCTAGATTTAGCGCTGTCCCCGCCGTC
GCACTACCCCAACTCTAACTTTAGGTTTAACTCGACGGGGCAGGCGGTGGCCGCCAAAAGCGTTGACTCTCTAGCCC
ACAACCTCTCCAAGGAGCCGTACCCCGCTCCTCTGCCGCCGCCCGTCTCCGGCGCCAACTCCACGGGCTCGGCCATC
CTCGACACCTACCTGCAGTTCATCACCGAGAACAGCCTCGGCATGGGCATGATGTCCTCGGAGCAGGCGGCGGCGGC
TGTGCACGCCGCCAAGCTGGCGCAACTCAACGCCATGGGCCTGGACAAGGCCTCGCAGCAGCACTTCTTGGAGAAGA
TGCAGCAGCAGATGCAGAGCGCGAACGAGCTCATGCTGAAGCGCAACGATGAGATACGGAGAGGCGATGCCGAGATG
GTGAATCATGACAGGGAGCGGGAGAAGGAAAATGACGAGGGGAGTCCTCGGGAGGCGAGGAGGAGGACTACAGTGA
GGAGGAAGCCGAACCCGAGGCGGTCAAAGCGGGAGAATAG
```

NCBI PREDICTED: Tribolium castaneum similar to abrupt CG4807-PA (LOC663820); (the cloned sequence is marked in yellow):

```
ATGGGGGGCGACACCTCCCCCGAGCAGCAGTACTCGCTGCGTTGGAATGACTTCCACTCGTCGATCCTCTCCTCGTT
TCGACACTTGAGGGACGAAGAAGATTTCGTCGATGTGACTTTGGCCTGTGATGGATGTTCGTTTACGGCACACAAAG
TTGTGCTGTCGGCGTGCAGTCCCTACTTTAGGAGGTTGCTCAAAGCCAATCCCTGCCAGCATCCCATCGTCATCCTC
AGGGACGTACAGCAAAAAGACATGGAAAGTCTGTTACGATTTATGTACAACGGAGAAGTTCACATAGGACAAGAACA
ATTAACGGATTTTTTAAAGACTGCACAAATGTTACAAGTGAGGGACTAGCCGACGTACCGGCGGGTACTGCAGGAC
```

Appendix

AAAGACTGACTGCTCCAGAGCAAAAGATCTCACCTAATTCTTCATCTTTACCGTGGGCGACGGACCGATCGGATGCT
CTCCGCGAGGGCGCTTTATCGCCGCCTCCGGCCAAAAGGTCCAGGAGTTCGGAGCGGGGGACGTACACCCCCGACAG
GGGCCGAAGTCCCTCCAGAGGCAACGGCGACATGCAGGAATCCCTCCTCGGACAGGCCTTGGAAGGCGGCCCCACGA
TACTCACAAAGGAGAAGGGAAATTCGGATTCGTCGCTGCAAGCGCAATCGACGGGCGAGGACAGCAACAGCAGCGAC
ACGGCGATGAGCGACCATGGAGACCCTGTAACCCCAAAGACTGAACCTTCCGACTACCCCATGCTAGACGACCACCA
TCCTTTCAATACCAACGGCGGGCTAATTGATCAGACGCGGACTCCGACATTTCCAGGCGCGCTGTTAGGTCTACAAG
GATTAGGAGGACTTATGCCAGGACCGTCAGGGATCCATAATAGTACGGACAATTTCGTTTCGAGACGCTCTTTGGAC
ATGATGAGAGTCAGAGCGACAGACCCAAGACCGTGTCCAAAGTGCGGCAAAATCTACCGATCTGCTCACACGTTACG
AACTCATTTAGAAGATAAACATACGATATGTCCAGGATATAGATGTGTTTTATGTGGAACTGTTGCGAAATCTCGAA
ATTCACTACATTCACATATGTCGCGACAACATAGAGGAATAAGTACAAAAGACCTGCCTGTTCTACCAATGCCCGCA
CCGTTCGACCCGGAACTTGCCAGCAGGTTGTTGGCAAAAGCCGGCGTGAAAATAAGTCCGGCCGAGCTGCGAGCGCG
CGCGTCTCCGACGGGTCCGCGTCGCTCCGACGTGAAGTTGGACGCGAAAAGCGCGTACTCTGGCGCGGCCTCGGAGG
CCGGCAGCTCAATATGCGGCGACAACGACCCCGAAGACCTGACAATCTCGCACCAAAGGTATGGAGGGCTAGATTTA
GCGCTGTCCCCGCCGTCGCACTACCCCAACTCTAACTTTAGGTTTAACTCGACGGGGCAGGCGGTGGCCGCCAAAAG
CGTTGACTCTCTAGCCCACAACCTCTCCAAGGAGCCGTACCCCGCTCCTCTGCCGCCGCCCGTCTCCGGCGCCAACT
CCACGGGCTCGGCCATCCTCGACACCTACCTGCAGTTCATCACCGAGAACAGCCTCGGCATGGGCATGATGTCCTCG
GAGCAGGCGGCGGCGGCTGTGCACGCCGCCAAGCTGGCGCAACTCAACGCCATGGGCCTGGACAAGGCCTCGCAGCA
GCACTTCTTGGAGAAGATGCAGCAGCAGATGCAGAGCGCGAACGAGCTCATGCTGAAGCGCAACGATGAGATACGGA
GAGGCGATGCCGAGATGGTGAATCATGACAGGGAGCGGGAGAAGGAAAATGACGAGGGGGAGTCCTCGGGAGGCGAG
GAGGAGGACTACAGTGAGGAGGAAGCCGAACCCGAGGCGGTCAAAGCGGGAGAATAG

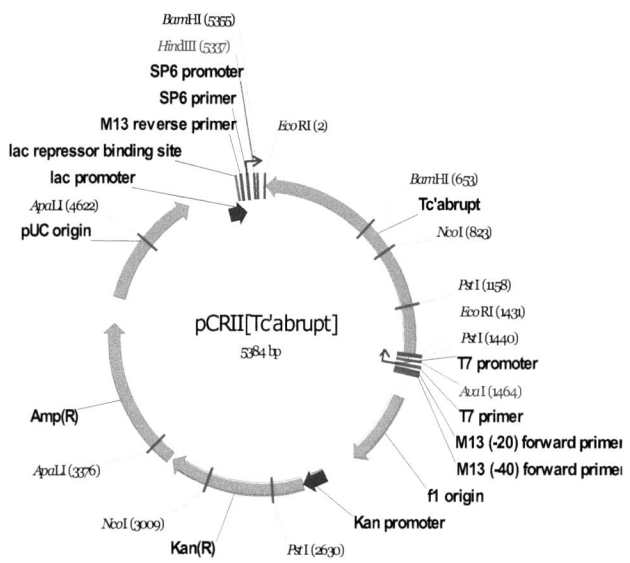

>pCRII[Tc'abrupt]

GAATTCGGCTTGGCGGCGGCAGAGGAGCGGGGTACGGCTCCTTGGAGAGGTTGTGGGCTAGAGAGTCAACGCTTTTG
GCGGCCACCGCCTGCCCCGTCGAATTAAACCTAAAGTTAGAGTTGGGGTAGTGCGACGGCGGGGACAGCGCTAAATC
TAGCCCTCCATACCTTTGGTGCGAGATTGTCAGGTCTTCGGGGTCGTTGTCGCCGCATATTGAGCTGCCGGCCTCCG
AGGCCGCGCCAGAGTACGCGCTTTTCGCGTCCAACTTCACGTCGGAGCGACGCGGACCCGTCGGAGACGCGCGCGCT
CGCAGCTCGGCCGGACTTATTTTCACGCCGGCTTTTGCCAACAACCTGCTGGCAAGTTCCGGGTCGAACGGTGCGGG
CATTGGTAGAACAGGCAGGTCTTTTGTACTTATTCCTCTATGTTGTCGCGACATATGTGAATGTAGTGAATTTCGAG
ATTTCGCAACAGTTCCACATAAAACACATCTATATCCTGGACATATCGTATGTTTATCTTCTAAATGAGTTCGTAAC
GTGTGAGCAGATCGGTAGATTTTGCCGCACTTTGGACACGGTCTTGGGTCTGTCGCTCTGACTCTCATCATGTCCAA
AGAGCGTCTCGAAAAGAAATTGTCCGTACTATTATGGATCCCTGACGGTCCTGGCATAAGTCCTCCTAATCCTTGTA

140

Appendix

```
GACCTAACAGCGCGCCTGGAAATGTCGGAGTCCGCGTCTGATCAATTAGCCCGCCGTTGGTATTGAAAGGATGGTGG
TCGTCTAGCATGGGGTAGTCGGAAGGTTCAGTCTTTGGGGTTACAGGGTCTCCATGGTCGCTCATCGCCGTGTCGCT
GCTGTTGCTGTCCTCGCCCGTCGATTGCGCTTGCAGCGACGAATCCGAATTTCCCTTCTCCTTTGTGAGTATCGTGG
GGCCGCCTTCCAAGGCCTGTCCGAGGAGGGATTCCTGCATGTCGCCGTTGCCTCTGGAGGGACTTCGGCCCCTGTCG
GGGGTGTACGTCCCCCGCTCCGAACTCCTGGACCTTTTGGCCGGAGGCGGCGATAAAGCGCCCTCGCGGAGAGCATC
CGATCGGTCCGTCGCCCACGGTAAAGATGAAGAATTAGGTGAGATCTTTTGCTCTGGAGCAGTCAGTCTTTGTCCTG
CAGTACCCGCCGGTACGTCGGCTAGTCCCCTCACTTGTAACATTTGTGCAGTCTTTAAAAAATCCGTTAATTGTTCT
TGTCCTATGTGAACTTCTCCGTTGTACATAAATCGTAACAGACTTTCCATGTCTTTTTGCTGTACGTCCCTGAGGAT
GACGATGGGATGCTGGCAGGGATTGGCTTTGAGCAACCTCCTAAAGTAGGGACTGCACGCCGACAGCACAACTTTGT
GTGCCGTAAACGAACATCCATCACAGGCCAAAGTCACAAAGCCGAATTCTGCAGATATCCATCACACTGGCGGCCGC
TCGAGCATGCATCTAGAGGGCCCAATTCGCCCTATAGTGAGTCGTATTACAATTCACTGGCCGTCGTTTTACAACGT
CGTGACTGGGAAAACCCTGGCGTTACCCAACTTAATCGCCTTGCAGCACATCCCCCTTTCGCCAGCTGGCGTAATAG
CGAAGAGGCCCGCACCGATCGCCCTTCCCAACAGTTGCGCAGCCTGAATGGCGAATGGACGCGCCCTGTAGCGGCGC
ATTAAGCGCGGCGGGTGTGGTGGTTACGCGCAGCGTGACCGCTACACTTGCCAGCGCCCTAGCGCCCGCTCCTTTCG
CTTTCTTCCCTTCCTTTCTCGCCACGTTCGCCGGCTTTCCCCGTCAAGCTCTAAATCGGGGGCTCCCTTTAGGGTTC
CGATTTAGTGCTTTACGGCACCTCGACCCCAAAAAACTTGATTAGGGTGATGGTTCACGTAGTGGGCCATCGCCCTG
ATAGACGGTTTTTCGCCCTTTGACGTTGGAGTCCACGTTCTTTAATAGTGGACTCTTGTTCCAAACTGGAACAACAC
TCAACCCTATCTCGGTCTATTCTTTTGATTTATAAGGGATTTTGCCGATTTCGGCCTATTGGTTAAAAAATGAGCTG
ATTTAACAAAAATTTAACGCGAATTTTAACAAAATTCAGGGCGCAAGGGCTGCTAAAGGAAGCGGAACACGTAGAAA
GCCAGTCCGCAGAAACGGTGCTGACCCCGGATGAATGTCAGCTACTGGGCTATCTGGACAAGGGAAAACGCAAGCGC
AAAGAGAAAGCAGGTAGCTTGCAGTGGGCTTACATGGCGATAGCTAGACTGGGCGGTTTTATGGACAGCAAGCGAAC
CGGAATTGCCAGCTGGGGCGCCCTCTGGTAAGGTTGGGAAGCCCTGCAAAGTAAACTGGATGGCTTTCTTGCCGCCA
AGGATCTGATGGCGCAGGGGATCAAGATCTGATCAAGAGACAGGATGAGGATCGTTTCGCATGATTGAACAAGATGG
ATTGCACGCAGGTTCTCCGGCCGCTTGGGTGGAGAGGCTATTCGGCTATGACTGGGCACAACAGACAATCGGCTGCT
CTGATGCCGCCGTGTTCCGGCTGTCAGCGCAGGGGCGCCCGGTTCTTTTTGTCAAGACCGACCTGTCCGGTGCCCTG
AATGAACTGCAGGACGAGGCAGCGCGGCTATCGTGGCTGGCCACGACGGGCGTTCCTTGCGCAGCTGTGCTCGACGT
TGTCACTGAAGCGGGAAGGGACTGGCTGCTATTGGGCGAAGTGCCGGGGCAGGATCTCCTGTCATCCCACCTTGCTC
CTGCCGAGAAAGTATCCATCATGGCTGATGCAATGCGGCGGCTGCATACGCTTGATCCGGCTACCTGCCCATTCGAC
CACCAAGCGAAACATCGCATCGAGCGAGCACGTACTCGGATGGAAGCCGGTCTTGTCGATCAGGATGATCTGGACGA
AGAGCATCAGGGGCTCGCGCCAGCCGAACTGTTCGCCAGGCTCAAGGCGCGCATGCCCGACGGCGAGGATCTCGTCG
TGACCCATGGCGATGCCTGCTTGCCGAATATCATGGTGGAAAATGGCCGCTTTTCTGGATTCATCGACTGTGGCCGG
CTGGGTGTGGCGGACCGCTATCAGGACATAGCGTTGGCTACCCGTGATATTGCTGAAGAGCTTGGCGGCGAATGGGC
TGACCGCTTCCTCGTGCTTTACGGTATCGCCGCTCCCGATTCGCAGCGCATCGCCTTCTATCGCCTTCTTGACGAGT
TCTTCTGAATTGAAAAAGGAAGAGTATGAGTATTCAACATTTCCGTGTCGCCCTTATTCCCTTTTTTGCGGCATTTT
GCCTTCCTGTTTTTGCTCACCCAGAAACGCTGGTGAAAGTAAAAGATGCTGAAGATCAGTTGGGTGCACGAGTGGGT
TACATCGAACTGGATCTCAACAGCGGTAAGATCCTTGAGAGTTTTCGCCCCGAAGAACGTTTTCCAATGATGAGCAC
TTTTAAAGTTCTGCTATGTGGCGCGGTATTATCCCGTATTGACGCCGGGCAAGAGCAACTCGGTCGCCGCATACACT
ATTCTCAGAATGACTTGGTTGAGTACTCACCAGTCACAGAAAAGCATCTTACGGATGGCATGACAGTAAGAGAATTA
TGCAGTGCTGCCATAACCATGAGTGATAACACTGCGGCCAACTTACTTCTGACAACGATCGGAGGACCGAAGGAGCT
AACCGCTTTTTTGCACAACATGGGGGATCATGTAACTCGCCTTGATCGTTGGGAACCGGAGCTGAATGAAGCCATAC
CAAACGACGAGCGTGACACCACGATGCCTGTAGCAATGGCAACAACGTTGCGCAAACTATTAACTGGCGAACTACTT
ACTCTAGCTTCCCGGCAACAATTAATAGACTGGATGGAGGCGGATAAAGTTGCAGGACCACTTCTGCGCTCGGCCCT
TCCGGCTGGCTGGTTTATTGCTGATAAATCTGGAGCCGGTGAGCGTGGGTCTCGCGGTATCATTGCAGCACTGGGGC
CAGATGGTAAGCCCTCCCGTATCGTAGTTATCTACACGACGGGGAGTCAGGCAACTATGGATGAACGAAATAGACAG
ATCGCTGAGATAGGTGCCTCACTGATTAAGCATTGGTAACTGTCAGACCAAGTTTACTCATATATACTTTAGATTGA
TTTAAAACTTCATTTTTAATTTAAAAGGATCTAGGTGAAGATCCTTTTTGATAATCTCATGACCAAAATCCCTTAAC
GTGAGTTTTCGTTCCACTGAGCGTCAGACCCCGTAGAAAAGATCAAAGGATCTTCTTGAGATCCTTTTTTTCTGCGC
GTAATCTGCTGCTTGCAAACAAAAAAACCACCGCTACCAGCGGTGGTTTGTTTGCCGGATCAAGAGCTACCAACTCT
TTTTCCGAAGGTAACTGGCTTCAGCAGAGCGCAGATACCAAATACTGTTCTTCTAGTGTAGCCGTAGTTAGGCCACC
ACTTCAAGAACTCTGTAGCACCGCCTACATACCTCGCTCTGCTAATCCTGTTACCAGTGGCTGCTGCCAGTGGCGAT
AAGTCGTGTCTTACCGGGTTGGACTCAAGACGATAGTTACCGGATAAGGCGCAGCGGTCGGGCTGAACGGGGGGTTC
GTGCACACAGCCCAGCTTGGAGCGAACGACCTACACCGAACTGAGATACCTACAGCGTGAGCTATGAGAAAGCGCCA
CGCTTCCCGAAGGGAGAAAGGCGGACAGGTATCCGGTAAGCGGCAGGGTCGGAACAGGAGAGCGCACGAGGGAGCTT
CCAGGGGGAAACGCCTGGTATCTTTATAGTCCTGTCGGGTTTCGCCACCTCTGACTTGAGCGTCGATTTTTGTGATG
CTCGTCAGGGGGCGGAGCCTATGGAAAAACGCCAGCAACGCGGCCTTTTTACGGTTCCTGGCCTTTTGCTGGCCTT
TTGCTCACATGTTCTTTCCTGCGTTATCCCCTGATTCTGTGGATAACCGTATTACCGCCTTTGAGTGAGCTGATACC
GCTCGCCGCAGCCGAACGACCGAGCGCAGCGAGTCAGTGAGCGAGGAAGCGGAAGAGCGCCCAATACGCAAACCGCC
TCTCCCCGCGCGTTGGCCGATTCATTAATGCAGCTGGCACGACAGGTTTCCCGACTGGAAAGCGGGCAGTGAGCGCA
ACGCAATTAATGTGAGTTAGCTCACTCATTAGGCACCCCAGGCTTTACACTTTATGCTTCCGGCTCGTATGTTGTGT
GGAATTGTGAGCGGATAACAATTTCACACAGGAAACAGCTATGACCATGATTACGCCAAGCTATTTAGGTGACACTA
TAGAATACTCAAGCTATGCATCAAGCTTGGTACCGAGCTCGGATCCACTAGTAACGGCCGCCAGTGTGCTG
```

Appendix

KS0294:

*piggy*Bac left arm flanking sequence:

TTAAATAAACTAGAATCTTGAGAACCAATAATAAACTGCTTTTTTCCCAATTTATTGCTTGATTTTTCTAATTTTTA
TTAGATTTTACACTTAGTTTCAAACTTTGTTATCGGGAACTGAAATCAATAGATAATTCTAAATAATTTCAAATTTT
AAATAAATGGGTTGACTTTTTGACATTTCGGACAAACCACAACAAAGTGTACTTACTAAACAAGGTGGATC

*piggy*Bac right arm flanking sequence:

TTAATATGTGTTATGAAAGAAATCGCTCTTTCAAATGCTTTTACAAAGCTGTCAAAATTAATTAATTCAATTAATCA
CAAATAGCACTTATGTAAATATCTCACACTATAATTTACACTGCAATTACGCTTATAATGATAGATGGAAACATCAA
TAACTGGTGAAACGTTTTGTTGGGGAAGAAATTGAATTGTTGGTCCACAGCCTAAGGCAAGAAAATTAACCTTGGGC
GTGTAATTGATCAAACAAACGCGAGATACCGGAAGTACTGAAAAACAGTCGCTCCAGGCCAGTGGGAACATCGATGT
TTTGTTTTGACGGACCCCTTACTCTCGTCTCATATAAACCGAAGCCAGCTAAGATGGTATACTTATTATCATCTTGT
GATGAGGATGCTTCTATCAACGAAAGTACCGGTAAACCGCAAATGGTTATGTATTATAATCAAACTAAAGGCGGAGT
GGACACGCTAGACCAAATGTGTTCTGTGATGACCTGCAGTAGGAAGACGAATAGGTGGCCTATGGCATTATTGTACG
GAATGATAAACATTGCCTGCATAAATTCTTTTATTATATACAGCCATAATGTCAGTAGCAAGGGAGAAAAGGTTCAA
AGTCGCAAAAAATTTATGAGAAACCTTTACATGAGCCTGACGTCATCGTTTATGCGTAAGCGTTTAGAAGCTCCTAC
TTTGAAGAGATATTTGCGC

Appendix

6.3 Sequences of GAL4/UAS constructs

pBac[3xP3-ECFP; Tc-hsp-GAL4Δ]

complement 1081..2127 pBacR

2136..2224 3xP3

2378..3097 ECFP

3108..3330 SV40polyA

complement 3402..3696 Tc-hsp68-3'UTR

complement 3980..4772 Gal4Delta

complement 4788..5485 Tc-hsp-5'

5643..6365 pBacL

SQ   Sequence 8723 BP; 2497 A; 1907 C; 1838 G; 2481 t;

```
     tcgcgcgttt cggtgatgac ggtgaaaacc tctgacacat gcagctcccg gagacggtca      60
     cagcttgtct gtaagcggat gccgggagca gacaagcccg tcagggcgcg tcagcgggtg     120
     ttggcgggtg tcggggctgg cttaactatg cggcatcaga gcagattgta ctgagagtgc     180
     accatatgcg gtgtgaaata ccgcacagat gcgtaaggag aaataccgc atcaggcgcc       240
     attcgccatt caggctgcgc aactgttggg aagggcgatc ggtgcgggcc tcttcgctat     300
     tacgccagct ggcgaaaggg ggatgtgctg caaggcgatt aagttgggta acgccagggt     360
     tttcccagtc acgacgttgt aaaacgacgg ccagtgccaa gctttgttta aaatataaca     420
     aaattgtgat cccacaaaat gaagtggggc aaaatcaaat aattaactag tgtccgtaaa     480
     cttgttggtc ttcaactttt tgaggaacac gttggacggc aaatcgtgac tataacacaa     540
     gttgatttaa taattttagc caacacgtcg ggctgcgtgt ttttgcgct ctgtgtacac       600
     gttgattaac tggtcgatta aataatttaa tttttggttc ttctttaaat ctgtgatgaa     660
     atttttttaaa ataactttaa attcttcatt ggtaaaaaat gccacgtttt gcaacttgtg    720
     agggtctaat atgaggtcaa actcagtagg agttttatcc aaaaagaaa acatgattac      780
     gtctgtacac gaacgcgtat taacgcagag tgcaaagtat aagagggtta aaaatatat     840
     tttacgcacc atatacgcat cgggttgata tcgttaatat ggatcaattt gaacagttga     900
     ttaacgtgtc tctgctcaag tctttgatca aacgcaaat cgacgaaaat gtgtcggaca       960
     atatcaagtc gatgagcgaa aaactaaaaa ggctagaata cgacaatctc acagacagcg    1020
     ttgagatata cggtattcac gacagcaggc tgaataataa aaaaattaga aactattatt    1080
     taaccctaga aagataatca tattgtgacg tacgttaaag ataatcatgc gtaaaattga    1140
     cgcatgtgtt ttatcggtct gtatatcgag gtttatttat taatttgaat agatattaag    1200
     ttttattata tttcacttta catactaata ataaattcaa caaacaattt atttatgttt    1260
     atttatttat taaaaaaaaa caaaaactca aaatttcttc tataaagtaa caaaactttt    1320
     aaacattctc tcttttacaa aaataaactt attttgtact ttaaaaacag tcatgttgta    1380
     ttataaaata agtaattagc ttaacttata cataatagaa acaaattata cttattagtc    1440
```

143

Appendix

```
agtcagaaac aactttggca catatcaata ttatgctctc gacaaataac tttttgcat   1500
tttttgcacg atgcatttgc ctttcgcctt attttagagg ggcagtaagt acagtaagta  1560
cgttttttca ttactggctc ttcagtactg tcatctgatg taccaggcac ttcatttggc  1620
aaaatattag agatattatc gcgcaaatat ctcttcaaag taggagcttc taaacgctta  1680
cgcataaacg atgacgtcag gctcatgtaa aggtttctca taaattttt gcgactttga   1740
accttttctc ccttgctact gacattatgg ctgtatataa taaaagaatt tatgcaggca  1800
atgtttatca ttccgtacaa taatgccata ggccacctat tcgtcttcct actgcaggtc  1860
atcacagaac acatttggtc tagcgtgtcc actccgcctt tagtttgatt ataatacata  1920
accatttgcg gtttaccggt actttcgttg atagaagcat cctcatcaca agatgataat  1980
aagtatacca tcttagctgg cttcggttta tatgagacga gagtaagggg tccgtcaaaa  2040
caaaacatcg atgttcccac tggcctggag cgactgtttt tcagtacttc cggtatctcg  2100
cgtttgtttg atcgcacggt tcccacaatg cggggattat tcattagaga ctaattcaat  2160
tagagctaat tcaattagga tccaagctta tcgatttcga accctcgacc gccggagtat  2220
aaatagaggc gcttcgtcta cggagcgaca attcaattca aacaagcaaa gtgaacacgt  2280
cgctaagcga aagctaagca aataaacaag cgcagctgaa caagctaaac aatcggggta  2340
ccgctagagt cgacggtacg atccaccggt cgccaccatg gtgagcaagg gcgaggagct  2400
gttcaccggg gtggtgccca tcctggtcga gctggacggc gacgtaaacg gccacaagtt  2460
cagcgtgtcc ggcgagggcg agggcgatgc cacctacggc aagctgaccc tgaagttcat  2520
ctgcaccacc ggcaagctgc ccgtgccctg gcccaccctc gtgaccaccc tgacctgggg  2580
cgtgcagtgc ttcagccgct accccgacca catgaagcag cacgacttct tcaagtccgc  2640
catgcccgaa ggctacgtcc aggagcgcac catcttcttc aaggacgacg gcaactacaa  2700
gacccgcgcc gaggtgaagt tcgagggcga caccctggtg aaccgcatcg agctgaaggg  2760
catcgacttc aaggaggacg gcaacatcct ggggcacaag ctggagtaca actacatcag  2820
ccacaacgtc tatatcaccg ccgacaagca gaagaacggc atcaaggcca acttcaagat  2880
ccgccacaac atcgaggacg gcagcgtgca gctcgccgac cactaccagc agaacacccc  2940
catcggcgac ggccccgtgc tgctgcccga caaccactac ctgagcaccc agtccgccct  3000
gagcaaagac cccaacgaga agcgcgatca catggtcctg ctggagttcg tgaccgccgc  3060
cgggatcact ctcggcatgg acgagctgta caagtaaagc ggccgcgact ctagatcata  3120
atcagccata ccacatttgt agaggtttta cttgctttaa aaaacctccc acacttcccc  3180
ctgaacctga aacataaaat gaatgcaatt gttgttgtta acttgtttat tgcagcttat  3240
aatggttaca aataaagcaa tagcatcaca aatttcacaa ataaagcatt tttttcactg  3300
cattctagtt gtggtttgtc caaactcatc atgtatctta agcttatcga tacgcgtacg  3360
gcgcgccaag cttaaggtgc acgcccacg tggccactag tgagcagcca gttgttttt   3420
tctatattcc tgatgggcat acgaaggaga gcccgtttgc gcttaatact gccaaggcca  3480
cgactaatgc ggtaagtttt ttattatttt aaaatgaaac aacttttat ctttaaatta   3540
aataatttat taaaattttt taaataatct acaaaaatca gtctttaaac ttaatgcagt  3600
cactatctac tttattataa ctaaagtaac tttacgtact attatttaca aaaataaact  3660
```

Appendix

```
atcacaatta actaagtagg aacaaaaacc gatcatctcg agctctgtac atgtccgcgg   3720
tcgaacgtct ctctagaagc ttctgaataa gccctcgtaa tataattttc tgagatttag   3780
gtccaaaaaa agatgggcat taattctagt catttaaaaa attctataga tcagaggtta   3840
catggccaag attgaaactt agaggagtat agttacataa aagaaggcaa aacgatgtat   3900
aaatgaaaga aattgagatg gtgcacgatg cacagttgaa gtgaacttgc ggggtttttc   3960
agtatctacg attcatttta ctctttttt gggtttggtg gggtatcttc atcatcgaat    4020
agatagttat atacatcatc cattgtagtg gtattaaaca tccctgtagt gattccaaac   4080
gcgttatacg cagtttggtc cgtccaacca ggtgacagtg gttttgaatt attaccatca   4140
tcaattttac tagccgtgat ttcattattc atgaagttat catgaacgtt agaggaggca   4200
attggttgtg aaagcgcttg agaatttgtt tgagttgtta tgaggttcgg accgttgcta   4260
ctgttagtga aagtgaagga caatgagcta tcagcaatat tcccactttg attaaaattg   4320
gcgaattccg gcgatacagt caactgtctt tgacctttgt tactactctc ttccgatgat   4380
gatgtcgcac ttattctatg ctgtctcaat gttagaggca tatcagtctc cactgaagcc   4440
aatctatctg tgacggcatc tttattcaca ttatcttgta caaataatcc tgttaacaat   4500
gcttttatat cctgtaaaga atccattttc aaaatcatgt caaggtcttc tcgaggaaaa   4560
atcagtagaa atagctgttc cagtctttct agccttgatt ccacttctgt cagatgtgcc   4620
ctagtcagcg gagacctttt ggttttggga gagtagcgac actcccagtt gttcttcaga   4680
cacttggcgc acttcggttt ttctttggag cacttgagct ttttaagtcg gcaaatatcg   4740
catgcttgtt cgatagaaga cagtagcttc atgggagacg tgtacctcta gatacacttt   4800
tcgatttact ttgaattcac tagtaaataa ttcactcaac tttgttaaag tcgctttgaa   4860
aatttgcttt gagtcgcttg cttagctttg ttgctttcgc ttcgcttgat tcaaattcac   4920
tgacaacgcg ccgcgagacc gcgcttatat atgaaacggg gaattcgtct cgaacttcac   4980
cgaatgatct cgaagtttcc attgatttgc gtttgaattt tctcgaattt ttatttttaa   5040
aacggccttt tcggtattat tcatgctatt tccgtaaaga taaaggagaa tagtggaaac   5100
tggttgtgcc ctaatttatt tcaacttcca aaataaaaac aatcattata tattaatgca   5160
tttaaaaaaa aaacgttaca taacgaatga gttgttgact atttcatgtt tatttttaat   5220
ttggtaaagt tcggcaatat tatgaatgag tagtgaattg ttttgatttc atatacatat   5280
gccaaagttg gattgtcata gcaattttaa taaaacaatt ttcctatttt ttttcttttt   5340
aaacttatta ataggtcttt aagtaaaatg ttttagatta atccttttt ttacttttat    5400
cgtttataat caattacttt tgcagaagat gcctaaataa tagtgattaa taataaataa   5460
agaacaattt tttattgaag gttgggatcc atatataggg cccgggttat aattacctca   5520
ggtcgacgtc ccatggccat tcgaattcgg ccggccgaat tcgaatggcc atgggacgtc   5580
gacctgaggt aattataacc cgggccctat atatggatcc aattgcaatg atcatcatga   5640
cagatctgac aatgttcagt gcagagactc ggctacgcct cgtggactt gaagttgacc    5700
aacaatgttt attcttacct ctaatagtcc tctgtggcaa ggtcaagatt ctgttagaag   5760
ccaatgaaga acctggttgt tcaataacat tttgttcgtc taatatttca ctaccgcttg   5820
acgttggctg cacttcatgt acctcatcta taaacgcttc ttctgtatcg ctctggacgt   5880
```

Appendix

```
catcttcact tacgtgatct gatatttcac tgtcagaatc ctcaccaaca agctcgtcat    5940

cgctttgcag aagagcagag aggatatgct catcgtctaa agaactaccc attttattat    6000

atattagtca cgatatctat aacaagaaaa tatatatata ataagttatc acgtaagtag    6060

aacatgaaat aacaatataa ttatcgtatg agttaaatct taaaagtcac gtaaaagata    6120

atcatgcgtc attttgactc acgcggtcgt tatagttcaa aatcagtgac acttaccgca    6180

ttgacaagca cgcctcacgg gagctccaag cggcgactga gatgtcctaa atgcacagcg    6240

acggattcgc gctatttaga aagagagagc aatatttcaa gaatgcatgc gtcaatttta    6300

cgcagactat ctttctaggg ttaaaaaaga tttgcgcttt actcgaccta aactttaaac    6360

acgtcataga atcttcgttt gacaaaaacc acattgtggc caagctgtgt gacgcgacgc    6420

gcgctaaaga atggcaaacc aagtcgcgcg agcgtcgact ctagaggatc cccgggtacc    6480

gagctcgaat tcgtaatcat ggtcatagct gtttcctgtg tgaaattgtt atccgctcac    6540

aattccacac aacatacgag ccggaagcat aaagtgtaaa gcctggggtg cctaatgagt    6600

gagctaactc acattaattg cgttgcgctc actgcccgct ttccagtcgg gaaacctgtc    6660

gtgccagctg cattaatgaa tcggccaacg cgcggggaga ggcggtttgc gtattgggcg    6720

ctcttccgct tcctcgctca ctgactcgct gcgctcggtc gttcggctgc ggcgagcggt    6780

atcagctcac tcaaaggcgg taatacggtt atccacagaa tcaggggata acgcaggaaa    6840

gaacatgtga gcaaaaggcc agcaaaaggc caggaaccgt aaaaaggccg cgttgctggc    6900

gtttttccat aggctccgcc ccctgacga gcatcacaaa aatcgacgct caagtcagag    6960

gtggcgaaac ccgacaggac tataaagata ccaggcgttt ccccctggaa gctccctcgt    7020

gcgctctcct gttccgaccc tgccgcttac cggatacctg tccgcctttc tcccttcggg    7080

aagcgtggcg ctttctcaat gctcacgctg taggtatctc agttcggtgt aggtcgttcg    7140

ctccaagctg ggctgtgtgc acgaaccccc cgttcagccc gaccgctgcg ccttatccgg    7200

taactatcgt cttgagtcca acccggtaag acacgactta tcgccactgg cagcagccac    7260

tggtaacagg attagcagag cgaggtatgt aggcggtgct acagagttct tgaagtggtg    7320

gcctaactac ggctacacta aaggacagt atttggtatc tgcgctctgc tgaagccagt    7380

taccttcgga aaaagagttg gtagctcttg atccggcaaa caaaccaccg ctggtagcgg    7440

tggtttttt gtttgcaagc agcagattac gcgcagaaaa aaaggatctc aagaagatcc    7500

tttgatcttt tctacggggt ctgacgctca gtggaacgaa aactcacgtt aagggatttt    7560

ggtcatgaga ttatcaaaaa ggatcttcac ctagatcctt ttaaattaaa aatgaagttt    7620

taaatcaatc taaagtatat atgagtaaac ttggtctgac agttaccaat gcttaatcag    7680

tgaggcacct atctcagcga tctgtctatt tcgttcatcc atagttgcct gactccccgt    7740

cgtgtagata actacgatac gggagggctt accatctggc cccagtgctg caatgatacc    7800

gcgagaccca cgctcaccgg ctccagattt atcagcaata aaccagccag ccggaagggc    7860

cgagcgcaga agtggtcctg caactttatc cgcctccatc cagtctatta attgttgccg    7920

ggaagctaga gtaagtagtt cgccagttaa tagtttgcgc aacgttgttg ccattgctac    7980

aggcatcgtg gtgtcacgct cgtcgtttgg tatggcttca ttcagctccg gttcccaacg    8040

atcaaggcga gttacatgat cccccatgtt gtgcaaaaaa gcggttagct ccttcggtcc    8100
```

```
tccgatcgtt gtcagaagta agttggccgc agtgttatca ctcatggtta tggcagcact    8160
gcataattct cttactgtca tgccatccgt aagatgcttt tctgtgactg gtgagtactc    8220
aaccaagtca ttctgagaat agtgtatgcg gcgaccgagt tgctcttgcc cggcgtcaat    8280
acgggataat accgcgccac atagcagaac tttaaaagtg ctcatcattg gaaaacgttc    8340
ttcggggcga aaactctcaa ggatcttacc gctgttgaga tccagttcga tgtaacccac    8400
tcgtgcaccc aactgatctt cagcatcttt tactttcacc agcgtttctg ggtgagcaaa    8460
aacaggaagg caaaatgccg caaaaaaggg aataagggcg acacggaaat gttgaatact    8520
catactcttc cttttcaat attattgaag catttatcag ggttattgtc tcatgagcgg    8580
atacatattt gaatgtattt agaaaaataa acaaatagg gttccgcgca catttccccg    8640
aaaagtgcca cctgacgtct aagaaaccat tattatcatg acattaacct ataaaaatag    8700
gcgtatcacg aggccctttc gtc    8723
```
//

Appendix

pBac[3xP3-EGFP; Tc-hsp-GAL4-VP16]

complement 1081..2127 pBacR

2136..2224 3xP3

2389..3108 EGFP

3119..3341 SV40polyA

complement 3413..3707 Tc'hsp-3'UTR

complement 3709..4459 Gal4-VP16

complement 4466..5163 Tc'hsp5'

5321..6043 pBacL

SQ   Sequence 8401 BP; 2363 A; 1903 C; 1795 G; 2340 t;

```
     tcgcgcgttt cggtgatgac ggtgaaaacc tctgacacat gcagctcccg gagacggtca      60
     cagcttgtct gtaagcggat gccgggagca gacaagcccg tcagggcgcg tcagcgggtg     120
     ttggcgggtg tcggggctgg cttaactatg cggcatcaga gcagattgta ctgagagtgc     180
     accatatgcg gtgtgaaata ccgcacagat gcgtaaggag aaaataccgc atcaggcgcc     240
     attcgccatt caggctgcgc aactgttggg aagggcgatc ggtgcgggcc tcttcgctat     300
     tacgccagct ggcgaaaggg ggatgtgctg caaggcgatt aagttgggta acgccagggt     360
     tttcccagtc acgacgttgt aaaacgacgg ccagtgccaa gctttgttta aaatataaca     420
     aaattgtgat cccacaaaat gaagtggggc aaaatcaaat aattaactag tgtccgtaaa     480
     cttgttggtc ttcaactttt tgaggaacac gttggacggc aaatcgtgac tataacacaa     540
     gttgatttaa taattttagc caacacgtcg ggctgcgtgt tttttgcgct ctgtgtacac     600
     gttgattaac tggtcgatta ataaatttaa ttttttggttc ttctttaaat ctgtgatgaa     660
     atttttttaaa ataactttaa attcttcatt ggtaaaaaat gccacgtttt gcaacttgtg     720
     agggtctaat atgaggtcaa actcagtagg agttttatcc aaaaaagaaa acatgattac     780
     gtctgtacac gaacgcgtat taacgcagag tgcaaagtat aagagggtta aaaaatatat     840
     tttacgcacc atatacgcat cgggttgata tcgttaatat ggatcaattt gaacagttga     900
     ttaacgtgtc tctgctcaag tctttgatca aaacgcaaat cgacgaaaat gtgtcggaca     960
     atatcaagtc gatgagcgaa aaactaaaaa ggctagaata cgacaatctc acagacagcg    1020
     ttgagatata cggtattcac gacagcaggc tgaataataa aaaaattaga aactattatt    1080
     taaccctaga aagataatca tattgtgacg tacgttaaag ataatcatgc gtaaaattga    1140
     cgcatgtgtt ttatcggtct gtatatcgag gtttatttat taatttgaat agatattaag    1200
     ttttattata tttacactta catactaata ataaattcaa caaacaattt atttatgttt    1260
     atttatttat taaaaaaaaa caaaaactca aaatttcttc tataaagtaa caaaactttt    1320
     aaacattctc tcttttacaa aaataaactt attttgtact ttaaaaacag tcatgttgta    1380
     ttataaaata agtaattagc ttaacttata cataatagaa acaaattata cttattagtc    1440
     agtcagaaac aactttggca catatcaata ttatgctctc gacaaataac tttttttgcat   1500
```

Appendix

```
tttttgcacg atgcatttgc ctttcgcctt attttagagg ggcagtaagt acagtaagta   1560
cgtttttttca ttactggctc ttcagtactg tcatctgatg taccaggcac ttcatttggc   1620
aaaatattag agatattatc gcgcaaatat ctcttcaaag taggagcttc taaacgctta   1680
cgcataaacg atgacgtcag gctcatgtaa aggtttctca taaattttttt gcgactttga   1740
accttttctc ccttgctact gacattatg ctgtatataa taaaagaatt tatgcaggca    1800
atgtttatca ttccgtacaa taatgccata ggccacctat tcgtcttcct actgcaggtc   1860
atcacagaac acatttggtc tagcgtgtcc actccgcctt tagtttgatt ataatacata   1920
accatttgcg gtttaccggt actttcgttg atagaagcat cctcatcaca agatgataat   1980
aagtatacca tcttagctgg cttcggttta tatgagacga gagtaagggg tccgtcaaaa   2040
caaaacatcg atgttcccac tggcctggag cgactgtttt tcagtacttc cggtatctcg   2100
cgtttgtttg atcgcacggt tcccacaatg cggggattat tcattagaga ctaattcaat   2160
tagagctaat tcaattagga tccaagctta tcgatttcga accctcgacc gccggagtat   2220
aaatagaggc gcttcgtcta cggagcgaca attcaattca aacaagcaaa gtgaacacgt   2280
cgctaagcga aagctaagca aataaacaag cgcagctgaa caagctaaac aatcggggta   2340
ccgctagagt cgacggtacc gcgggcccgg gatccaccgg tcgccaccat ggtgagcaag   2400
ggcgaggagc tgttcaccgg ggtggtgccc atcctggtcg agctggacgg cgacgtaaac   2460
ggccacaagt tcagcgtgtc cggcgagggc gagggcgatg ccacctacgg caagctgacc   2520
ctgaagttca tctgcaccac cggcaagctg cccgtgccct ggcccaccct cgtgaccacc   2580
ctgacctacg gcgtgcagtg cttcagccgc taccccgacc acatgaagca gcacgacttc   2640
ttcaagtccg ccatgcccga aggctacgtc caggagcgca ccatcttctt caaggacgac   2700
ggcaactaca agacccgcgc cgaggtgaag ttcgagggcg acaccctggt gaaccgcatc   2760
gagctgaagg gcatcgactt caaggaggac ggcaacatcc tggggcacaa gctggagtac   2820
aactacaaca gccacaacgt ctatatcatg gccgacaagc agaagaacgg catcaaggtg   2880
aacttcaaga tccgccacaa catcgaggac ggcagcgtgc agctcgccga ccactaccag   2940
cagaacaccc ccatcggcga cggccccgtg ctgctgcccg acaaccacta cctgagcacc   3000
cagtccgccc tgagcaaaga ccccaacgag aagcgcgatc acatggtcct gctggagttc   3060
gtgaccgccg ccgggatcac tctcggcatg gacgagctgt acaagtaaag cggccgcgac   3120
tctagatcat aatcagccat accacatttg tagaggtttt acttgctttaa aaaacctcc   3180
cacacttccc cctgaacctg aaacataaaa tgaatgcaat tgttgttgtt aacttgttta   3240
ttgcagctta taatggttac aaataaagca atagcatcac aaatttcaca aataaagcat   3300
ttttttcact gcattctagt tgtggtttgt ccaaactcat catgtatctt aagcttatcg   3360
atacgcgtac ggcgcgccaa gcttaaggtg cacggcccac gtggccacta gtgagcagcc   3420
agttgttttt ttctatattc ctgatgggca tacgaaggag agcccgtttg cgcttaatac   3480
tgccaaggcc acgactaatg cggtaagttc tttattattt taaaatgaaa caacttttta   3540
tctttaaatt aaataattta ttaaaatttt ttaaataatc tacaaaaatc agtctttaaa   3600
```

149

Appendix

```
cttaatgcag tcactatcta ctttattata actaaagtaa ctttacgtac tattatttac    3660
aaaaataaac tatcacaatt aactaagtag gaacacaaac cgatcatctc gagctctgta    3720
catgtccgcg gtcgtagtag atccgctaca tatccagagc gccgtagggg gcggagtcgt    3780
gggggtaaa tcccggaccc ggggaatccc cgtccccaa catgtccaga tcgaaatcgt     3840
ctagcgcgtc ggcatgcgcc atcgccacgt cctcgccgtc taagtggagc tcgtccccca    3900
ggctgacatc ggtcggggg gcggctcgag agatctgaat tcccggggtc gacctcgacg    3960
atacagtcaa ctgtctttga cctttgttac tactctcttc cgatgatgat gtcgcactta    4020
ttctatgctg tctcaatgtt agaggcatat cagtctccac tgaagccaat ctatctgtga    4080
cggcatcttt attcacatta tcttgtacaa ataatcctgt taacaatgct tttatatcct    4140
gtaaagaatc cattttcaaa atcatgtcaa ggtcttctcg aggaaaaatc agtagaaata    4200
gctgttccag tctttctagc cttgattcca cttctgtcag atgtgcccta gtcagcggag    4260
accttttggt tttgggagag tagcgacact cccagttgtt cttcagacac ttggcgcact    4320
tcggtttttc tttggagcac ttgagctttt taagtcggca aatatcgcat gcttgttcga    4380
tagaagacag tagcttcatc tttcaggagg cttgcttcaa gctccttgaa ttcgaatcga    4440
tgggatctcg actctagcgg tacctctaga tgcacttttc gatttacttt gaattcacta    4500
gtaaataatt cactcaactt tgttaaagtc gctttgaaaa tttgctttga gtcgcttgct    4560
tagctttgtt gctttcgctt cgcttgattc aaattcactg acaacgcgcc gcgagaccgc    4620
gcttatatat gaaacgggga attcgtctcg aacttcaccg aatgatctcg aagtttccat    4680
tgatttgcgt ttgaattttc tcgaattttt attttaaaa cggccttttc ggtattattc    4740
atgctatttc cgtaaagata aaggagaata gtggaaactg gttgtgccct aatttatttc    4800
aacttccaaa ataaaaacaa tcattatata ttaatgcatt taaaaaaaaa acgttacata    4860
acgaatgagt tgttgactat ttcatgttta tttttaattt ggtaaagttc ggcaatatta    4920
tgaatgagta gtgaattgtt ttgatttcat atacatatgc caaagttgga ttgtcatagc    4980
aattttaata aaacaatttt cctatttttt ttcttttaa acttattaat aggtctttaa    5040
gtaaaatgtt ttagattaat cctttttttt acttttatcg tttataatca attacttttg    5100
cagaagatgc ctaaataata gtgattaata ataataaag aacaattttt tattgaaggt    5160
tgggatccat atatagggcc cgggttataa ttacctcagg tcgacgtccc atggccattc    5220
gaattcggcc ggccgaattc gaatggccat gggacgtcga cctgaggtaa ttataacccg    5280
ggccctatat atggatccaa ttgcaatgat catcatgaca gatctgacaa tgttcagtgc    5340
agagactcgg ctacgcctcg tggactttga agttgaccaa caatgtttat tcttacctct    5400
aatagtcctc tgtggcaagg tcaagattct gttagaagcc aatgaagaac ctggttgttc    5460
aataacattt tgttcgtcta atatttcact accgcttgac gttggctgca cttcatgtac    5520
ctcatctata aacgcttctt ctgtatcgct ctggacgtca tcttcactta cgtgatctga    5580
tatttcactg tcagaatcct caccaacaag ctcgtcatcg ctttgcagaa gagcagagag    5640
gatatgctca tcgtctaaag aactacccat tttattatat attagtcacg atatctataa    5700
```

Appendix

```
caagaaaata tatatataat aagttatcac gtaagtagaa catgaaataa caatataatt      5760

atcgtatgag ttaaatctta aaagtcacgt aaaagataat catgcgtcat tttgactcac      5820

gcggtcgtta tagttcaaaa tcagtgacac ttaccgcatt gacaagcacg cctcacggga      5880

gctccaagcg gcgactgaga tgtcctaaat gcacagcgac ggattcgcgc tatttagaaa      5940

gagagagcaa tatttcaaga atgcatgcgt caattttacg cagactatct ttctagggtt      6000

aaaaaagatt tgcgctttac tcgacctaaa ctttaaacac gtcatagaat cttcgtttga      6060

caaaaaccac attgtggcca agctgtgtga cgcgacgcgc gctaaagaat ggcaaaccaa      6120

gtcgcgcgag cgtcgactct agaggatccc cgggtaccga gctcgaattc gtaatcatgg      6180

tcatagctgt ttcctgtgtg aaattgttat ccgctcacaa ttccacacaa catacgagcc      6240

ggaagcataa agtgtaaagc ctggggtgcc taatgagtga gctaactcac attaattgcg      6300

ttgcgctcac tgcccgcttt ccagtcggga aacctgtcgt gccagctgca ttaatgaatc      6360

ggccaacgcg cggggagagg cggtttgcgt attgggcgct cttccgcttc ctcgctcact      6420

gactcgctgc gctcggtcgt tcggctgcgg cgagcggtat cagctcactc aaaggcggta      6480

atacggttat ccacagaatc aggggataac gcaggaaaga acatgtgagc aaaaggccag      6540

caaaaggcca ggaaccgtaa aaaggccgcg ttgctggcgt ttttccatag gctccgcccc      6600

cctgacgagc atcacaaaaa tcgacgctca agtcagaggt ggcgaaaccc gacaggacta      6660

taaagatacc aggcgtttcc ccctggaagc tccctcgtgc gctctcctgt tccgaccctg      6720

ccgcttaccg gatacctgtc cgcctttctc ccttcgggaa gcgtggcgct ttctcaatgc      6780

tcacgctgta ggtatctcag ttcggtgtag gtcgttcgct ccaagctggg ctgtgtgcac      6840

gaaccccccg ttcagcccga ccgctgcgcc ttatccggta actatcgtct tgagtccaac      6900

ccggtaagac acgacttatc gccactggca gcagccactg gtaacaggat tagcagagcg      6960

aggtatgtag gcggtgctac agagttcttg aagtggtggc ctaactacgg ctacactaga      7020

aggacagtat ttggtatctg cgctctgctg aagccagtta ccttcggaaa aagagttggt      7080

agctcttgat ccggcaaaca aaccaccgct ggtagcggtg gtttttttgt ttgcaagcag      7140

cagattacgc gcagaaaaaa aggatctcaa gaagatcctt tgatcttttc tacgggtct      7200

gacgctcagt ggaacgaaaa ctcacgttaa gggattttgg tcatgagatt atcaaaaagg      7260

atcttcacct agatcctttt aaattaaaaa tgaagtttta atcaatcta aagtatatat      7320

gagtaaactt ggtctgacag ttaccaatgc ttaatcagtg aggcacctat ctcagcgatc      7380

tgtctatttc gttcatccat agttgcctga ctccccgtcg tgtagataac tacgatacgg      7440

gagggcttac catctggccc cagtgctgca atgataccgc gagacccacg ctcaccggct      7500

ccagatttat cagcaataaa ccagccagcc ggaagggccg agcgcagaag tggtcctgca      7560

actttatccg cctccatcca gtctattaat tgttgccggg aagctagagt aagtagttcg      7620

ccagttaata gtttgcgcaa cgttgttgcc attgctacag gcatcgtggt gtcacgctcg      7680

tcgtttggta tggcttcatt cagctccggt tcccaacgat caaggcgagt tacatgatcc      7740

cccatgttgt gcaaaaaagc ggttagctcc ttcggtcctc cgatcgttgt cagaagtaag      7800
```

Appendix

```
ttggccgcag tgttatcact catggttatg gcagcactgc ataattctct tactgtcatg      7860

ccatccgtaa gatgcttttc tgtgactggt gagtactcaa ccaagtcatt ctgagaatag      7920

tgtatgcggc gaccgagttg ctcttgcccg gcgtcaatac gggataatac cgcgccacat      7980

agcagaactt taaaagtgct catcattgga aaacgttctt cggggcgaaa actctcaagg      8040

atcttaccgc tgttgagatc cagttcgatg taacccactc gtgcacccaa ctgatcttca      8100

gcatctttta ctttcaccag cgtttctggg tgagcaaaaa caggaaggca aaatgccgca      8160

aaaaagggaa taagggcgac acggaaatgt tgaatactca tactcttcct ttttcaatat      8220

tattgaagca tttatcaggg ttattgtctc atgagcggat acatatttga atgtatttag      8280

aaaaataaac aaatagaggt tccgcgcaca tttccccgaa aagtgccacc tgacgtctaa      8340

gaaaccatta ttatcatgac attaacctat aaaaataggc gtatcacgag gccctttcgt      8400

c                                                                     8401
//
```

Appendix

pBac[3xP3-DsRed; UAS-Tc-bhsp-tGFP]

1081..2127   pBacR

2136..2224   3xP3

2389..3069   DsRed

3078..3300   SV40polyA

3449..3580   5xUAS

3584..3734   Tc-hsp_p

3751..4449   turboGFP

4449..4653   SV40polyA

4724..5446   pBacL

SQ   Sequence 7804 BP; 2139 A; 1898 C; 1791 G; 1976 t;

tcgcgcgttt	cggtgatgac	ggtgaaaacc	tctgacacat	gcagctcccg	gagacggtca	60
cagcttgtct	gtaagcggat	gccgggagca	gacaagcccg	tcagggcgcg	tcagcgggtg	120
ttggcgggtg	tcggggctgg	cttaactatg	cggcatcaga	gcagattgta	ctgagagtgc	180
accatatgcg	gtgtgaaata	ccgcacagat	gcgtaaggag	aaaataccgc	atcaggcgcc	240
attcgccatt	caggctgcgc	aactgttggg	aagggcgatc	ggtgcgggcc	tcttcgctat	300
tacgccagct	ggcgaaaggg	ggatgtgctg	caaggcgatt	aagttgggta	acgccagggt	360
tttcccagtc	acgacgttgt	aaaacgacgg	ccagtgccaa	gctttgttta	aatataaca	420
aaattgtgat	cccacaaaat	gaagtgggc	aaatcaaat	aattaactag	tgtccgtaaa	480
cttgttggtc	ttcaacttt	tgaggaacac	gttggacggc	aaatcgtgac	tataacacaa	540
gttgatttaa	taattttagc	caacacgtcg	ggctgcgtgt	ttttgcgct	ctgtgtacac	600
gttgattaac	tggtcgatta	aataatttaa	ttttttggttc	ttctttaaat	ctgtgatgaa	660
atttttaaa	ataactttaa	attcttcatt	ggtaaaaaat	gccacgtttt	gcaacttgtg	720
agggtctaat	atgaggtcaa	actcagtagg	agttttatcc	aaaaaagaaa	acatgattac	780
gtctgtacac	gaacgcgtat	taacgcagag	tgcaaagtat	aagagggtta	aaaaatatat	840
tttacgcacc	atatacgcat	cgggttgata	tcgttaatat	ggatcaattt	gaacagttga	900
ttaacgtgtc	tctgctcaag	tctttgatca	aaacgcaaat	cgacgaaaat	gtgtcggaca	960
atatcaagtc	gatgagcgaa	aaactaaaaa	ggctagaata	cgacaatctc	acagacagcg	1020
ttgagatata	cggtattcac	gacagcaggc	tgaataataa	aaaaattaga	aactattatt	1080
taaccctaga	aagataatca	tattgtgacg	tacgttaaag	ataatcatgc	gtaaaattga	1140
cgcatgtgtt	ttatcggtct	gtatatcgag	gtttatttat	taatttgaat	agatattaag	1200
ttttattata	tttacactta	catactaata	ataaattcaa	caaacaattt	atttatgttt	1260
atttattat	taaaaaaaaa	caaaaactca	aaatttcttc	tataaagtaa	caaaacttt	1320
aaacattctc	tcttttacaa	aaataaactt	attttgtact	ttaaaaacag	tcatgttgta	1380
ttataaaata	agtaattagc	ttaacttata	cataatagaa	acaaattata	cttattagtc	1440
agtcagaaac	aactttggca	catatcaata	ttatgctctc	gacaaataac	tttttgcat	1500

153

Appendix

```
tttttgcacg atgcatttgc ctttcgcctt attttagagg ggcagtaagt acagtaagta   1560

cgtttttca ttactggctc ttcagtactg tcatctgatg taccaggcac ttcatttggc    1620

aaaatattag agatattatc gcgcaaatat ctcttcaaag taggagcttc taaacgctta   1680

cgcataaacg atgacgtcag gctcatgtaa aggtttctca taaattttt gcgactttga    1740

accttttctc ccttgctact gacattatgg ctgtatataa taaaagaatt tatgcaggca   1800

atgtttatca ttccgtacaa taatgccata ggccacctat tcgtcttcct actgcaggtc   1860

atcacagaac acatttggtc tagcgtgtcc actccgcctt tagtttgatt ataatacata   1920

accatttgcg gtttaccggt actttcgttg atagaagcat cctcatcaca agatgataat   1980

aagtatacca tcttagctgg cttcggttta tatgagacga gagtaagggg tccgtcaaaa   2040

caaaacatcg atgttccccac tggcctggag cgactgtttt tcagtacttc cggtatctcg   2100

cgtttgtttg atcgcacggt tcccacaatg cggggattat tcattagaga ctaattcaat   2160

tagagctaat tcaattagga tccaagctta tcgatttcga accctcgacc gccggagtat   2220

aaatagaggc gcttcgtcta cggagcgaca attcaattca aacaagcaaa gtgaacacgt   2280

cgctaagcga aagctaagca aataaacaag cgcagctgaa caagctaaac aatcggggta   2340

ccgctagagt cgacggtacc gcgggcccgg gatccaccgg tcgccaccat ggtgcgctcc   2400

tccaagaacg tcatcaagga gttcatgcgc ttcaaggtgc gcatggaggg caccgtgaac   2460

ggccacgagt tcgagatcga gggcgagggc gagggccgcc cctacgaggg ccacaacacc   2520

gtgaagctga aggtgaccaa gggcggcccc ctgcccttcg cctgggacat cctgtccccc   2580

cagttccagt acggctccaa ggtgtacgtg aagcaccccg ccgacatccc cgactacaag   2640

aagctgtcct tccccgaggg cttcaagtgg gagcgcgtga tgaacttcga ggacggcggc   2700

gtggtgaccg tgacccagga ctcctccctg caggacggct gcttcatcta caaggtgaag   2760

ttcatcggcg tgaacttccc ctccgacggc cccgtaatgc agaagaagac catgggctgg   2820

gaggcctcca ccgagcgcct gtacccccgc gacggcgtgc tgaagggcga gatccacaag   2880

gccctgaagc tgaaggacgg cggccactac ctggtggagt tcaagtccat ctacatggcc   2940

aagaagcccg tgcagctgcc cggctactac tacgtggact ccaagctgga catcacctcc   3000

cacaacgagg attacaccat cgtggagcag tacgagcgca ccgagggccg ccaccacctg   3060

ttcctgtagc ggccgcgact ctagatcata atcagccata ccacatttgt agaggtttta   3120

cttgctttaa aaaacctccc acacttcccc ctgaacctga aacataaaat gaatgcaatt   3180

gttgttgtta acttgtttat tgcagcttat aatggttaca aataaagcaa tagcatcaca   3240

aatttcacaa ataaagcatt ttttcactg cattctagtt gtggtttgtc caaactcatc    3300

atgtatctta agcttatcga tacgcgtacg gcgcgcctag gccggccgaa ttcgaatggc   3360

catgggacgt cgacctgagg taattataac ccgggcccta tatggatc caattgcaat    3420

gatcatcatg acagatctgc gcgcgatcga tatctgcagg tcggagtact gtcctccgag   3480

cggagtactg tcctccgagc ggagtactgt cctccgagcg gagtactgtc ctccgagcgg   3540

agtactgtcc tccgagcgga gactctagcg agcgccggag atccgtttca tataagcg    3600

cggtctcgcg gcgcgttgtc agtgaatttg aatcaagcga agcgaaagca acaaagctaa   3660

gcaagcgact caaagcaaat tttcaaagcg actttaacaa agttgagtga attatttact   3720
```

154

Appendix

```
agtgaattca aagtggtacc ggtcgccacc atggagagcg acgagagcgg cctgcccgcc     3780

atggagatcg agtgccgcat caccggcacc ctgaacggcg tggagttcga gctggtgggc     3840

ggcggagagg gcaccccega gcagggccgc atgaccaaca agatgaagag caccaaaggc     3900

gccctgacct tcagcccta cctgctgagc cacgtgatgg gctacggctt ctaccacttc     3960

ggcacctacc ccagcggcta cgagaacccc ttcctgcacg ccatcaacaa cggcggctac     4020

accaacaccc gcatcgagaa gtacgaggac ggcggcgtgc tgcacgtgag cttcagctac     4080

cgctacgagg ccggccgcgt gatcggcgac ttcaaggtga tgggcaccgg cttcccegag     4140

gacagcgtga tcttcaccga caagatcatc cgcagcaacg ccaccgtgga gcacctgcac     4200

cccatgggcg ataacgatct ggatggcagc ttcacccgca ccttcagcct gcgcgacggc     4260

ggctactaca gctccgtggt ggacagccac atgcacttca agagcgccat ccacccagc     4320

atcctgcaga acggggccc catgttcgcc ttccgccgcg tggaggagga tcacagcaac     4380

accgagctgg gcatcgtgga gtaccagcac gccttcaaga ccccggatgc agatgccggt     4440

gaagaataaa gcggccgcga ctctagatca taatcagcca taccacattt gtagaggttt     4500

tacttgcttt aaaaaacctc ccacacctcc ccctgaacct gaaacataaa atgaatgcaa     4560

ttgttgttgt taacttgttt attgcagctt ataatggtta caaataaagc aatagcatca     4620

caaatttcac aaataaagca tttttttcac tgcattctag ttgtggtttg tccaaactca     4680

tcaatgtatc ttaagcttgg cgcgccagg ccggccgatc tcggatctga caatgttcag     4740

tgcagagact cggctacgcc tcgtggactt tgaagttgac caacaatgtt tattcttacc     4800

tctaatagtc ctctgtggca aggtcaagat tctgttagaa gccaatgaag aacctggttg     4860

ttcaataaca ttttgttcgt ctaatatttc actaccgctt gacgttggct gcacttcatg     4920

tacctcatct ataaacgctt cttctgtatc gctctggacg tcatcttcac ttacgtgatc     4980

tgatatttca ctgtcagaat cctcaccaac aagctcgtca tcgctttgca gaagagcaga     5040

gaggatatgc tcatcgtcta aagaactacc cattttatta tatattagtc acgatatcta     5100

taacaagaaa atatatatat aataagttat cacgtaagta gaacatgaaa taacaatata     5160

attatcgtat gagttaaatc ttaaaagtca cgtaaaagat aatcatgcgt cattttgact     5220

cacgcggtcg ttatagttca aaatcagtga cacttaccgc attgacaagc acgcctcacg     5280

ggagctccaa gcggcgactg agatgtccta aatgcacagc gacggattcg cgctatttag     5340

aaagagagag caatatttca agaatgcatg cgtcaatttt acgcagacta tctttctagg     5400

gttaaaaaag atttgcgctt tactcgacct aaactttaaa cacgtcatag aatcttcgtt     5460

tgacaaaaac cacattgtgg ccaagctgtg tgacgcgacg cgcgctaaag aatggcaaac     5520

caagtcgcgc gagcgtcgac tctagaggat ccccgggtac cgagctcgaa ttcgtaatca     5580

tggtcatagc tgtttcctgt gtgaaattgt tatccgctca caattccaca caacatacga     5640

gccggaagca taaagtgtaa agcctgggg gcctaatgag tgagctaact cacattaatt     5700

gcgttgcgct cactgcccgc tttccagtcg ggaaacctgt cgtgccagct gcattaatga     5760

atcggccaac gcgcggggag aggcggtttg cgtattgggc gctcttccgc ttcctcgctc     5820

actgactcgc tgcgctcggt cgttcggctg cggcgagcgg tatcagctca ctcaaaggcg     5880

gtaatacggt tatccacaga atcaggggat aacgcaggaa agaacatgtg agcaaaaggc     5940
```

155

Appendix

```
cagcaaaagg ccaggaaccg taaaaaggcc gcgttgctgg cgtttttcca taggctccgc    6000
cccctgacg agcatcacaa aaatcgacgc tcaagtcaga ggtggcgaaa cccgacagga     6060
ctataaagat accaggcgtt tccccctgga agctccctcg tgcgctctcc tgttccgacc    6120
ctgccgctta ccggatacct gtccgccttt ctcccttcgg gaagcgtggc gctttctcaa   6180
tgctcacgct gtaggtatct cagttcggtg taggtcgttc gctccaagct gggctgtgtg   6240
cacgaacccc ccgttcagcc cgaccgctgc gccttatccg gtaactatcg tcttgagtcc   6300
aacccggtaa gacacgactt atcgccactg gcagcagcca ctggtaacag gattagcaga   6360
gcgaggtatg taggcggtgc tacagagttc ttgaagtggt ggcctaacta cggctacact   6420
agaaggacag tatttggtat ctgcgctctg ctgaagccag ttaccttcgg aaaaagagtt   6480
ggtagctctt gatccggcaa acaaaccacc gctggtagcg gtggtttttt tgtttgcaag   6540
cagcagatta cgcgcagaaa aaaaggatct caagaagatc ctttgatctt ttctacgggg   6600
tctgacgctc agtggaacga aaactcacgt taagggattt tggtcatgag attatcaaaa   6660
aggatcttca cctagatcct tttaaattaa aaatgaagtt ttaaatcaat ctaaagtata   6720
tatgagtaaa cttggtctga cagttaccaa tgcttaatca gtgaggcacc tatctcagcg   6780
atctgtctat ttcgttcatc catagttgcc tgactccccg tcgtgtagat aactacgata   6840
cgggagggct taccatctgg ccccagtgct gcaatgatac cgcgagaccc acgctcaccg   6900
gctccagatt tatcagcaat aaaccagcca gccggaaggg ccgagcgcag aagtggtcct   6960
gcaactttat ccgcctccat ccagtctatt aattgttgcc gggaagctag agtaagtagt   7020
tcgccagtta atagtttgcg caacgttgtt gccattgcta caggcatcgt ggtgtcacgc   7080
tcgtcgtttg gtatggcttc attcagctcc ggttcccaac gatcaaggcg agttacatga   7140
tcccccatgt tgtgcaaaaa agcggttagc tccttcggtc ctccgatcgt tgtcagaagt   7200
aagttggccg cagtgttatc actcatggtt atggcagcac tgcataattc tcttactgtc   7260
atgccatccg taagatgctt ttctgtgact ggtgagtact caaccaagtc attctgagaa   7320
tagtgtatgc ggcgaccgag ttgctcttgc ccggcgtcaa tacgggataa taccgcgcca   7380
catagcagaa ctttaaaagt gctcatcatt ggaaaacgtt cttcggggcg aaaactctca   7440
aggatcttac cgctgttgag atccagttcg atgtaaccca ctcgtgcacc caactgatct   7500
tcagcatctt ttactttcac cagcgtttct gggtgagcaa aacaggaag gcaaaatgcc    7560
gcaaaaaagg gaataagggc gacacggaaa tgttgaatac tcatactctt cctttttcaa   7620
tattattgaa gcatttatca gggttattgt ctcatgagcg gatacatatt tgaatgtatt   7680
tagaaaaata aacaaatagg ggttccgcgc acatttcccc gaaaagtgcc acctgacgtc   7740
taagaaacca ttattatcat gacattaacc tataaaaata ggcgtatcac gaggcccttt   7800
cgtc                                                                7804
```

6.4 Abbreviations

RNAi	RNA interference
cDNA	complementary DNA
dsRNA	double-stranded RNA
mRNA	messenger RNA
iPCR	inverse PCR
tGFP	turbo green fluorescent protein
EGFP	enhanced green fluorescent protein
EYFP	enhanced yellow fluorescent protein
ECFP	enhanced cyan fluorescent protein
vw	vermilion white

Die VDM Verlagsservicegesellschaft sucht für wissenschaftliche Verlage abgeschlossene und herausragende

Dissertationen, Habilitationen, Diplomarbeiten, Master Theses, Magisterarbeiten usw.

für die kostenlose Publikation als Fachbuch.

Sie verfügen über eine Arbeit, die hohen inhaltlichen und formalen Ansprüchen genügt, und haben Interesse an einer honorarvergüteten Publikation?

Dann senden Sie bitte erste Informationen über sich und Ihre Arbeit per Email an *info@vdm-vsg.de*.

Sie erhalten kurzfristig unser Feedback!

VDM Verlagsservicegesellschaft mbH
Dudweiler Landstr. 99 Telefon +49 681 3720 174
D - 66123 Saarbrücken Fax   +49 681 3720 1749
www.vdm-vsg.de

Die VDM Verlagsservicegesellschaft mbH vertritt

Printed by Books on Demand GmbH, Norderstedt / Germany